T0318630

Development and Application of Classical Capillary Number Curve Theory

Development and Application of Classical Capillary Number Curve Theory

Lianqing Qi

Yanjun Yin

Yu Wang

Qigui Ma

Hongshen Wang

Gulf Professional Publishing
An imprint of Elsevier

Gulf Professional Publishing is an imprint of Elsevier
50 Hampshire Street, 5th Floor, Cambridge, MA 02139, United States
The Boulevard, Langford Lane, Kidlington, Oxford, OX5 1GB, United Kingdom

Library of Congress Cataloging-in-Publication Data
A catalog record for this book is available from the Library of Congress

British Library Cataloguing-in-Publication Data
A catalogue record for this book is available from the British Library

ISBN: 978-0-12-821225-7

For information on all Gulf Professional publications
visit our website at https://www.elsevier.com/books-and-journals

Publisher: Jonathan Simpson
Acquisitions Editor: Glyn Jones
Editorial Project Manager: Naomi Robertson
Production Project Manager: Poulouse Joseph
Cover Designer: Matthew Limbert

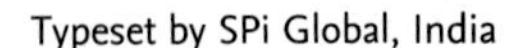
Typeset by SPi Global, India

Contents

Contributors

Numbers in parentheses indicate the pages on which the authors' contributions begin.

Shijia Chen (57), CNOOC Energy Technology & Services-Drilling & Production Technology Services Co., Tianjin, China

Wei He (123), CNOOC Energy Technology & Services-Drilling & Production Technology Services Co., Tianjin, China

Jirui Hou (1), Enhanced Oil Recovery Institute of China University of Petroleum, Beijing, China

Bo Huang (1), CNOOC Energy Technology & Services-Drilling & Production Technology Services Co., Tianjin, China

Bailing Kong (123), SINOPEC Henan Oilfield Exploration and Development Institute, Henan, China

DaoShan Li (91), CNPC Dagang Oilfield Production Technology Institute, Tianjin, China

Fang Li (91), CNOOC Energy Technology & Services-Drilling & Production Technology Services Co., Tianjin, China

Feng Li (57), CNOOC Energy Technology & Services-Drilling & Production Technology Services Co., Tianjin, China

Jianlu Li (163), CNPC Daqing Oilfield Exploration and Development Research Institute, Daqing, China

Ronghua Li (123), CNPC Xinjiang Oilfield Experiment and Detection Institute, Xinjiang, China

Yiqiang Li (193), Enhanced Oil Recovery Institute of China University of Petroleum, Beijing, China

Chuntian Liu (57), CNPC Daqing Oilfield Exploration and Development Research Institute, Daqing, China

Quangang Liu (31), CNOOC Energy Technology & Services-Drilling & Production Technology Services Co., Tianjin, China

Zongzhao Liu (1), China United Coalbed Methane Corporation, Ltd., Beijing, China

Ying Ma (123), CNPC Huabei Oilfield Company Petroleum Exploration and Production Research Institute, Renqiu, China

Lianqing Qi (1, 31, 57, 91, 123, 163, 193), CNPC Daqing Oilfield Exploration and Development Research Institute, Daqing; CNOOC Energy Technology & Services-Drilling & Production Technology Services Co., Tianjin, China

Weihong Qiao (91), Chemical Engineering College of Dalian University of Technology, Liaoning, China

Liantao Shan (123), SINOPEC Shengli Oilfield Exploration and Development Institute, Shandong, China

Si Shen (31), CNOOC Energy Technology & Services-Drilling & Production Technology Services Co., Tianjin, China

Duansheng Shi (123), CNOOC Energy Technology & Services-Drilling & Production Technology Services Co., Tianjin, China

Fenggang Shi (1), CNOOC Energy Technology & Services-Drilling & Production Technology Services Co., Tianjin, China

Yong Shi (193), CNOOC Energy Technology & Services-Safety & Environmental Protection Co., Tianjin, China

Daowan Song (163), SINOPEC Shengli Oilfield Exploration and Development Institute, Shandong, China

Kaoping Song (31, 163, 193), Northeast Petroleum University, Daqing, China

Shuai Tan (163), CNOOC Energy Technology & Services-Drilling & Production Technology Services Co., Tianjin, China

Chengsheng Wang (57), CNOOC Energy Technology & Services-Drilling & Production Technology Services Co., Tianjin, China

Hongshen Wang (31, 163), CNOOC Energy Technology & Services-Drilling & Production Technology Services Co., Tianjin, China

Jinlin Wang (31), CNOOC Energy Technology & Services-Drilling & Production Technology Services Co., Tianjin, China

Qiang Wang (123), CNPC Exploration and Development Institute, Beijing, China

Xiaochao Wang (31), CNOOC Energy Technology & Services-Drilling & Production Technology Services Co., Tianjin, China

Yu Wang (91), CNPC Xinjiang Oilfield Experiment and Detection Institute, Xinjiang, China

Jun Wei (193), CNOOC Energy Technology & Services-Drilling & Production Technology Services Co., Tianjin, China

Dali Weng (193), CNOOC Energy Technology & Services-Drilling & Production Technology Services Co., Tianjin, China

Shenqu Wu (163), CNOOC Energy Technology & Services-Drilling & Production Technology Services Co., Tianjin, China

Yali Wu (91), CNOOC Energy Technology & Services-Drilling & Production Technology Services Co., Tianjin, China

Yi Wu (193), CNPC, Liaohe Oilfield Exploration and Development Institute, Liaoning, China

Guanli Xu (57), SINOPEC Petroleum Exploration and Production Research Institute, Beijing, China

Chengzhi Yang (1), CNPC Research Institute of Exploration and Development, Beijing, China

Zhongbin Ye (57), School of Chemistry and Chemical Engineering of Southwest Petroleum University, Chengdu, China

Yanjun Yin (1, 57, 91), CNOOC Energy Technology & Services-Drilling & Production Technology Services Co., Tianjin, China

Tao Yu (31), CNPC Liaohe Oilfield Exploration and Development Institute, Panjin, China

Jian Zhang (1), CNOOC Research Institute, Beijing, China

Junhui Zhang (91), CNOOC Energy Technology & Services-Drilling & Production Technology Services Co., Tianjin, China

Ning Zhang (163), CNOOC China Ltd Tianjin Branch, Tianjin, China

Hongqing Zhu (193), CNOOC Energy Technology & Services-Drilling & Production Technology Services Co., Tianjin, China

Foreword

Mr. Lianqing Qi is an old friend of mine whom I became acquainted with in the research work of oil recovery enhancement. Since Mr. Qi came to Daqing Oilfield from Dalian, Liaoning Province in 1980, he has been engaged in the numerical simulation research of oil recovery enhancement by chemical flooding. This collection of papers is the summary of the results of chemical combination flooding by Mr. Qi and his team for the past 20 years.

The classic capillary number experimental curve was put forward by American scholars in the middle of the 20th century. The authors of this book, on the basis of studying and researching the American scholars' achievement, found a strange phenomenon of change on the capillary number curve. They found that when the capillary number was greater than the limit capillary number, the capillary number increased as did residual oil saturation. This makes it difficult to use the classical theory explanation in numerical simulation calculation. Over nearly a decade of concentrated study and experimentation, the "capillary number of the experimental curve *QL*" was achieved.

This book, based on the basic theory and definition of seepage mechanics, analyzes and studies experimental data of the capillary number experimental curve *QL*. The chemical oil displacement experiment core micro-spatial characteristics of the oil and water distribution model is established, which shows the structure of core microscopic pore space and the condition of oil and water distribution. The establishment of the microscopic model explains the capillary number curve morphologic change, deepening understanding of the capillary number curve and making the study of chemical flooding experiment (test) deep into reservoir microscopic space. Based on the theoretical research results, the authors put forward the "digital oil displacement test" research method, which was used to study chemical flooding in four oil production plants in Daqing Oilfield.

The authors studied the ultra-low interfacial tension system of Sho-Vel-Tum Oil Field in Oklahoma, United States. They carried out a "sensitivity" analysis on the slug structure of the flooding scheme, the system's interfacial tension, underground viscosity, injection velocity, and a number of other factors.

Through thousands of "digital oil displacement tests," the authors put forth a two-stage optimization scheme for Daqing Oilfield. The oil displacement scheme with high viscosity and ultra-low interfacial tension system aims to produce the crude oil originally accumulated in the microscopic pure oil and pure oil space V_O in the oil layer. The recovery factor can reach greater than 30% relative to water flooding. The oil displacement scheme with high viscosity and ultra-low interfacial tension system expands the recovery target to the microscopic oil-water coexistence space with smaller pore diameter, and the recovery rate can be increased by another 2%.

It is hoped that the publication of this book will enlighten researchers in the field and help them to apply the innovative results contained herein. The information provided will help to serve oilfield development and contributes to the research and application of chemical compound flooding technology.

Pingping Shen

Preface

In the middle of the 20th century, American scholars put forward the concept of the capillary number, and gave the curve of the relationship between the capillary number and residual oil through experiments. These experiments revealed that as the capillary number increases, the corresponding residual oil saturation decreases, and the residual oil saturation no longer decreases after reaching the limit of capillary number. This important achievement laid the theoretical foundation of enhanced oil recovery by chemical flooding. On this basis, the scholars created UTCHEM, a numerical simulation software model for chemical flooding. In the numerical simulation study of chemical combination flooding, I found that when the capillary number was higher than the limit capillary number, the capillary number increased and the residual oil saturation also increased. Meanwhile, the flooding experiment verified this situation. The "curve" of the "capillary number" calculation was obtained by numerical simulation, and the relatively complete description of the curve was given. On this basis, the experiment resulted in the "chemical combination flooding capillary number experiment curve *QL*," which is different from the classical capillary curve. Based on the "chemical combination flooding capillary number experiment curve *QL*," the corresponding "chemical combination flooding relative permeability curve *QL*" expression was written, and the corresponding relative permeability curve–related parameters were measured by experiments. This new research achievement sheds light on previous doubts and opens up new research horizons. Based on these innovative theory research results and together with a software expert, Mr. Jialin Dai developed the Improved Mechanism of Combination Flooding Simulation (IMCFS), a chemical flooding software. The capillary number "chemical combination flooding experiment curve *QL*" and "chemical combination flooding relative permeability curve *QL*" were written into the software so as to improve the description of the chemical combination flooding mechanism and provide effective technical support for the research and application of chemical combination flooding technology.

Based on the geological structure model matched with the chemical flooding simulation, using IMCFS software, the numerical simulation fitting calculation of field test was carried out in Xingerxi Block area of No. 4 Oil Production Plant in Daqing Oilfield. The field experiment was digitized and the digital geologic model platform containing reservoir geologic information and combination flooding information was established, on which the scheme calculation was carried out to optimize an oil displacement plan. The research method can be referred to as the "digital oil displacement experiment." Oilfields that use chemical combination flooding can adopt this research method. For oilfields that are suitable for chemical combination flooding but have not carried out oil displacement tests of chemical combination flooding, water flooding and chemical combination oil displacement experiments could be carried out on the three-dimensional core model made by combining with specific reservoir conditions. This experiment is equivalent to accomplishing a micro field test, and based on the micro field test, establishing a geological model of digital research platform, to carry out the "digital oil displacement experiment," accelerate the research and application of chemical flooding technology in the oilfield.

Based on the theory and definition of seepage mechanics, the characteristics model of oil and water distribution in the microcosmic space of oil displacement experiment core was established by analyzing the data of the capillary number experiment curve *QL*. Sorted by order of pore size from large to small: "pure oil" space V_O, "pure water" space V_W, "oil-water coexistence" space, and so on. This is an important research result that reveals the structure of microscopic pore space and the distribution of oil and water in the reservoir core. The limited capillary number N_{CC} on the experimental curve *QL* of the capillary number is the limited capillary number of water flooding and polymer flooding. The crude oil produced by water flooding and polymer flooding is the crude oil originally accumulated in V_{O1}, a subspace with relatively large voids in "pure oil" space V_O, and the ultimate value of residual oil saturation is V_{Or}^L. The capillary number is higher than that of N_{CC}, so it is converted to chemical combination flooding. The crude oil produced is the crude oil originally accumulated in the relatively small pore size subspace V_{O2} in "pure oil" space V_O. When the capillary number increases to the limit capillary number N_{ct2}, the residual oil saturation reaches the limit value S_{or}^H the capillary number continues to increase, displacement enters the "pure water" space V_W, and the residual oil no longer changes. When the capillary number is higher than the limit capillary number N_{ct2}, the displacement enters into the "oil-water coexistence" space. In the displacement process, there is the "pure oil" subspace V_W, in which the crude oil is released and extracted, and the "pure water" subspace V_{ow} captured flowing oil, which is converted into "bound oil," and the capillary number curve presents a complex change form. Setting up of microcosmic space oil and the water distribution model is an important contribution to reservoir geology, although the experimental study was done only on ordinary sandstone core, not on the core with cracks and holes, however, the results of the study can be applied on core with holes and cracks, because cracks and holes are pore space with a larger pore size, which is an important part of the "pure oil" spaces V_o. It should also be noted here that experimental research methods can be used to completely having cracks, holes on the experimental core, measured the corresponding experimental capillary number curve, deepening the understanding the research to the cores with holes and fractures. The construction of a micro-spatial oil and water distribution characteristics model has potential important application value in oilfield exploration and development.

By using the digital experimental research method, the "capillary number" digitization "experimental curve" composed of multiple curves was obtained, and it was confirmed that each capillary number curve was the corresponding relationship curve between the capillary number higher than N_C and residual oil saturation S_{or} under a certain displacement condition. When the capillary number is less than the limit capillary number N_{CC}, during the oil displacement, there is only one capillary number curve. When the capillary number is within the limit of capillary number between N_{CC} and N_{ct}, displacement is carried out on the different V/μ_w value of displacement conditions, resulting in an orderly arrangement of the capillary number curve. While the capillary number experimental curve QL is the envelope curve of the bunch curve, the classic capillary number is a curve corresponding to the experimental curve. Its displacement condition value of V/μ_w is relatively bigger and its limit experiment curve of capillary number N_{ct} lies between the limit capillary number N_{CC} and N_{ct} of the capillary number experiment curve QL. Under the condition that the capillary number is higher than the limit capillary number N_{CC}, and under the displacement condition with different V/μ_w values, a bunch of ordered capillary number curves is obtained. The practice of "digitalization" of the capillary number experimental curve deepens and improves the understanding of the capillary number experimental curve.

By applying the digital experimental research method, thousands of "oil displacement tests" have been conducted for chemical combination flooding field tests in Daqing Oilfield. And the "sensitivity analysis" method has been adopted to optimize and study various elements in the oil displacement schemes.

The oil displacement scheme of the ultra-low viscosity system aims to produce crude oil that is currently accumulated in V_O and V_W in the microscopic space of the oil reservoir. It can be achieved by adjusting and correcting the current oil field schemes. It has good operability.

The oil displacement scheme with a high viscosity and ultra-low interfacial tension system aims to produce the crude oil originally accumulated in the microscopic pure oil and pure oil space V_O and V_W in the oil reservoir. The recovery factor can reach >30%, increasing oil equivalent to 55 tons of polymer oil relative to water flooding. The oil displacement scheme with a high viscosity and ultra-low interfacial tension system expands the recovery target to the microscopic oil-water coexistence space with smaller pore diameter, and the recovery rate can be increased by another 2%. In the oil flooding scheme of a high viscosity and ultra-low interfacial tension system, the goal of oil recovery is expanded to the micro-oil-water coexistence space of the produced reservoir, and the recovery rate can be increased by another 2% compared with the high viscosity, ultra-low system oil flooding scheme.

This book consists of seven chapters that bring together the results of 18 years of hard work by my team and me. In it, we elaborate on the important achievements in chemical combination flooding technology and present innovations in the research process.

The application of chemical combination flooding technology in Daqing Oilfield is of great value. The successful experience of the oil displacement test is integrated with the results of our research team to achieve "high-quality" development and create perfect "Chinese technology" that can be applied to developing domestic oil fields, including Daqing Oilfield, Liaohe Oilfield, and Xinjiang Oilfield, as well as oil fields overseas. This book is full of innovative research results that we are convinced will attract the attention of domestic and foreign colleagues.

Lianqing Qi

Authors' biography

Lianqing Qi is a senior engineer and national science and technology project leader. He graduated from Dalian University of Technology in 1966, under the tutorship of academician Lingxi Qian (Prof Lingxi Qian, Chinese Academy of Sciences), as a senior engineer he joined the chemical flooding test and numerical simulation in Daqing Oilfield in 1980. In 2003, he was employed as a chemical oil displacement expert by China National Offshore Oil Company's (CNOOC) EnerTech-Drilling and Production Company and led the chemical flooding and enhance oil recovery (EOR) research team for 15 years, obtained a number of experimental research results, published monographs, and many research papers on chemical flooding.

Yanjun Yin is a senior reservoir engineer for the CNOOC EnerTech-Drilling and Production Company in Tianjin, China. He graduated from Northeast Petroleum University in 2004 with a bachelor degree and is mainly engaged in reservoir engineering scheme design, reservoir numerical simulation, and tertiary oil recovery technology research work.

Yu Wang is a researcher for the Xinjiang Oilfield Branch-Company of China National Petroleum Corporation. He graduated from Xinjiang Chemistry Institute of the Chinese Academy of Sciences in 1997 with a Master's degree in Organic Chemistry, and China Petroleum University (Beijing) in 2007 with a Doctor's degree in Chemical Engineering. He has been engaged in the research of enhanced oil recovery technology for a long time, and is especially good at the research of chemical flooding formula design. He has participated in the research of cheap formulae of chemical flooding and compound flooding technology of the conglomerate reservoir in Xinjiang.

Qigui Ma is a member of the China Translation Association, and senior engineer. He graduated from Heilongjiang University in 1967, and in 1973 worked in the Research Institute of Petroleum and Exploration, at Daqing Oilfield. He has been engaged in Petroleum Science and Technology Intelligence Translation work both oral and written translation for 33 years. He twice went to France during 1984–85 to attend "Improve Oil Recovery Methods Selection Scheme" on research work, several times went to the United States to Improve Oil Recovery Scheme Review during 1986–87; and attended many international petroleum conferences, SPE Conferences, etc., as an interpreter. In 2012, he won the title of senior translator of the China Translation Association.

Hongshen Wang is a senior reservoir engineer for the CNOOC EnerTech-Drilling and Production Company in Tianjin, China. He graduated from China Petroleum University (Beijing) in 2002 with a Master's degree and is mainly engaged in reservoir engineering scheme design, and tertiary oil recovery technology research work.

Informative abstracts

Classic capillary number experimental curve is the theoretical basis of chemical oil displacement. However, there are some shortcomings in its practical application. Through experiments presented in this volume, the authors created a "chemical compound displacement capillary experiment curve *QL*," which supplements and improves on the classic capillary curve. Based on the basic theory and definition of percolation mechanics, the authors analyzed and studied data of experimental curve *QL* and established the oil and water distribution characteristics model of microscopic space in the core of oil displacement. They wrote the results of this theoretical innovation into numerical simulation software and applied them to the simulation calculation, thus creating the "digital oil displacement test" method.

This book is a useful reference for petroleum researchers, oilfield engineers, and teachers and students at petroleum colleges and universities.

Afterword

Dear friends, thank you for reading this book. You should now have a better understanding of chemical oil displacement technology, from theory to application of the technology.

At present, the international crude oil price is low, and there is little pressure on oil fields' production. This makes for a great opportunity for scientific research and public relations. We hope this book inspires researchers to study and design combination flooding optimization schemes, apply them to oilfield development, and extract precious underground resources as efficiently as possible. Let us work together to open a new chapter in the research and application of chemical combination flooding technology.

We would like to express our gratitude to three academicians at the Chinese Academy of Engineering (CAE)—Mr. Demin Wang, Dakuang Han, and Gengdong Cheng—and Professors Pingping Shen, Chengzhi Yang, and Youyi Zhu. We also wish to thank Chief Engineer Qimin Wang at Daqing Oilfield, and Chief Engineer of the Research Institute of Exploration and Development of Daqing Oilfield, Xiaolin Wu. Finally, thanks go to software experts Jialin Dai and Weidong Cao of Shengli Oilfield for their care and support for this research.

Chapter 1

Supplement and optimization of classical capillary number experimental curve for enhanced oil recovery by combination flooding

Lianqing Qi[a,b], Zongzhao Liu[c], Chengzhi Yang[d], Yanjun Yin[b], Jirui Hou[e], Jian Zhang[f], Bo Huang[b], and Fenggang Shi[b]

[a]*CNPC Daqing Oilfield Exploration and Development Research Institute, Daqing, China,* [b]*CNOOC Energy Technology & Services-Drilling & Production Technology Services Co., Tianjin, China,* [c]*China United Coalbed Methane Corporation, Ltd., Beijing, China,* [d]*CNPC Research Institute of Exploration and Development, Beijing, China,* [e]*Enhanced Oil Recovery Institute of China University of Petroleum, Beijing, China,* [f]*CNOOC Research Institute, Beijing, China*

Chapter outline

Development and Application of Classical Capillary Number Curve Theory
https://doi.org/10.1016/B978-0-12-821225-7.00001-1

1.1 Classical capillary number experimental curve and chemical flooding numerical simulation software UTCHEM

In the middle of last century, in order to study and describe "the relations between the hydrodynamic force and capillary retention force which make the trapped residual oil to move" during oil displacement, Moore and Slobod,[1] Taber,[2] and Foster[3] put forward a concept of a ratio of hydrodynamic force to capillary force, called capillary number. Its definition formula is

$$N_c = \frac{V\mu_w}{\sigma_{ow}}, \tag{1.1}$$

where N_c is capillary number (dimensionless), V is displacing phase seepage velocity (m/s), μ_w is displacing phase viscosity (mPa s), and σ_{ow} is interfacial tension between displacing phase and displaced phase (mN/m). The corresponding relation curve between capillary number and residual oil is given through further experiments, commonly called "capillary number curve." Scholars obtain curves with different shapes from different angles. Fig. 1.1 shows an experimental curve finished by Moore and Slobod.

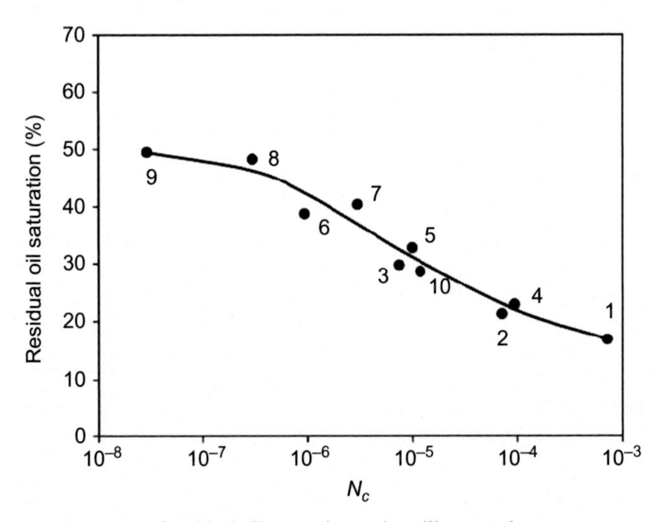

FIG. 1.1 Relation curve of residual oil saturation and capillary number.

For aiding the laboratory experimental studies, in some software (e.g., FACS), which is developed from the famous chemical flooding numerical simulation software UTCHEM made by scholars of the University of Texas (UT), the mathematic expression of capillary number curve is[4]:

$$S_{lr} = S_{lr}^{H} + \frac{S_{lr}^{L} - S_{lr}^{H}}{1 + T_l N_{cl}}, \quad l = w, o, \tag{1.2}$$

Where T_l is a constant, S_{lr}^{L}, S_{lr}^{H} are remaining saturation of liquid phase (l) at low capillary number and ideal high capillary number, respectively.

The change of remaining saturation of phase will cause change in relative permeability curve, and the relative permeability curve of phase l after change is expressed as:

$$K_{rl} = K_{rl}^{0} (S_{nl})^{n_l}, \quad l = w, o, \tag{1.3}$$

where S_{nl} is a "normalized" saturation of phase l:

$$S_{nl} = \frac{S_l - S_{lr}}{1 - S_{lr} - S_{l'r}}, \quad l = w, o, \; l' = o, w. \tag{1.4}$$

The calculation expressions for the end point value k_{rl}^{0} and exponent n_l respectively are

$$K_{rl}^{0} = K_{rl}^{L} + \frac{S_{l'r}^{L} - S_{l'r}}{S_{l'r}^{L} - S_{l'r}^{H}} \left(K_{rl}^{H} - K_{rl}^{L}\right), \quad l = w, o, \; l' = o, w, \tag{1.5}$$

$$n_l = n_l^{L} + \frac{S_{rl'}^{L} - S_{rl'}}{S_{rl'}^{L} - S_{rl'}^{H}} \left(n_l^{H} - n_l^{L}\right), \quad l = w, o, \; l' = o, w, \tag{1.6}$$

Where K_{rl}^{L} is the end point value of relative permeability curve with low capillary number, n_l^{L} is exponent of relative permeability curve with low capillary number, K_{rl}^{H} is end point value of relative permeability curve with ideal capillary number, and n_l^{H} is the exponent of relative permeability curve with ideal capillary number.

The proposal of capillary number (N_c) and the obtainment of experimental curve for the relation between capillary number and residual oil saturation have laid a theoretical foundation for the study of combination flooding technology. Currently, laboratory experiments and field test studies are carried out in many countries, including China and the United states. The Daqing Oilfield in China has entered the industrial test stage. The development of UTCHEM software pushes forward the study and application of combination flooding technology.

1.2 Two different driving conditions found in the combination flooding process in numerical simulation study

Using the FACS software developed by US Grand Company, an in-depth study on simulation calculation is carried out in combination flooding field tests in the Xingerxi Test Area of Daqing Oilfield, China. The study shows that two different driving conditions exist in the combination flooding process, and through analysis with oil displacement experiments and field tests, it is determined that two different driving conditions exist in the oil displacement process. Qi et al.[5, 6] give a detailed description, and a brief introduction is given here.

1.2.1 Conditions and results for simulation calculation study

The geological models for simulation calculation are as follows:

- 3D model, $N_x = N_y = 9, N_z = 3$ (N_x, N_y, and N_z represent grid number in X, Y, and Z directions respectively), homogeneous in plane, heterogeneous in longitudinal direction, the permeability of three layers from top to bottom is 100, 210, and 525×10^{-3} μm^2
- 2D plane model, $N_x = N_y = 9$, $N_z = 1$, permeability 450×10^{-3} μm^2
- 1D model, $N_x = 9$, $N_y = N_z = 1$, permeability 450×10^{-3} μm^2

With reference to the geological and fluid characteristics of Daqing Oilfield, data related to geology and fluid required by simulation calculation are determined.

In the calculation study, the termination condition for oil displacement schemes is that wells have a comprehensive water cut of 98%. In alkaline-surfactant-polymer (ASP) combination flooding schemes, the recovery percentage of reserves at which water cut decreases and then returns to 98% again during combination flooding is defined as the recovery percentage of reserves for ASP flooding, and the difference between it and the recovery percentage of reserves at 98% water cut in corresponding water flooding schemes is defined as the increased range of oil recovery factor for APS flooding.

Change law with the same oil displacement effect is found in the calculation with 3D, 2D, and 1D models. Under low capillary number condition during the oil displacement process, oil displacement effect improves with the increase of capillary number, and the absorption ratio of surfactant (the percentage of absorbed surfactant volume to the total injected volume at the end of the scheme) is relatively small. This is called "Type I" driving condition. When the capillary number in the oil layer exceeds a certain limit capillary number during oil displacement, it turns into a "Type II" driving condition where oil displacement effect becomes poor. The pressure in the oil layer increases, water phase velocity increases, and absorption ratio of surfactant increases significantly. Further, with the increase of system viscosity and capillary number, oil displacement effect improves again because a high-viscosity system inhibits water phase.

Table 1.1 gives the calculation results for the 1D model. The fluid injection speed in the "extra super-low" scheme is taken as 0.15 PV/a, and the system interfacial tension is taken as 1×10^{-4} mN/m. In the "extra high-speed, extra super-low" scheme, extra high-fluid injection speed is taken as 0.3 PV/a, with an extra super-low system interfacial tension. R in the table is the percentage of reserve recovery of the scheme, and Ads is the surfactant absorption ratio.

TABLE 1.1 Corresponding data at the end of oil displacement scheme in 1D model under certain conditions.

		System viscosity μ (mPa s)							
Scheme type	**Item**	***6.19***	***9.80***	***14.0***	***18.4***	***23.8***	***30.5***	***38.5***	***46.3***
Extra super low	R (%)	88.27	92.66	94.59	95.61	96.17	96.65	96.98	85.66
	Ads (%)	0.18	0.17	0.15	0.13	0.15	0.20	0.20	6.17
	N_c	0.019	0.032	0.046	0.061	0.079	0.102	0.129	0.140
	S_{or}	0.096	0.060	0.044	0.036	0.031	0.027	0.025	0.117
Extra high speed, extra super low	R (%)	91.06	94.56	96.05	96.72	86.27	86.30	87.49	89.01
	Ads (%)	0.32	0.31	0.25	0.16	5.71	5.88	6.45	6.48
	N_c	0.039	0.064	0.093	0.123	0.145	0.186	0.237	0.289
	S_{or}	0.073	0.044	0.032	0.027	0.112	0.112	0.102	0.090

Based on the results in the table, in the "extra super-low" scheme, oil displacement effect changes dramatically when system viscosity is higher than 38.5 mPa s. Meanwhile, in the "extra high-speed, extra super-low" scheme, oil displacement effect changes dramatically when system viscosity is higher than 18.4 mPa s. It is noted that the calculation schemes here terminate at water cut of 98% in produced fluid, and the change of remaining oil saturation can roughly reflect the change of residual oil saturation. Therefore, the two groups of data are converted to the relations between "residual" oil saturation and capillary number and included in Table 1.1.

Through analysis of the relations between capillary number and residual oil saturation in the table, it can be seen that there is a limit value N_{ct} between capillary numbers 0.129 and 0.140; when the capillary number is less than the limit capillary number N_{ct}, the residual oil saturation decreases with the increase of capillary number, and the limit capillary number N_{ct} corresponds to the minimum value of the residual oil saturation. After that, with the increase of system viscosity and capillary number, the residual oil saturation increases first, and then decreases.

It can be seen by contrast that the classical capillary number experimental curve in Fig. 1.1 does not show complex changes between capillary number and residual oil saturation at capillary number exceeding the limit value.

1.2.2 Verification with oil displacement experiment

The corresponding relation between capillary number and residual oil saturation is obtained through verification with oil displacement experiment. Considering that the seepage velocity is high in the high-permeability layer in the 2D section model, two different driving conditions tend to occur in oil displacement experiments. Therefore, batches of oil displacement experiments are conducted using the 2D section model. The size of artificial cores used in experiments is 4 cm × 4 cm × 30 cm, and cores are divided into three layers with the same thickness in longitudinal direction, with each layer homogeneous in plane and heterogeneous in longitudinal direction. Their permeability are 260, 710, and 1200×10^{-3} μm^2 respectively from top to bottom. Also considering that the oil displacement experiment with 1D homogenous core can provide more precise capillary number in oil displacement process, two groups of 1D core oil displacement experiments are done, with core size of 4 × 4 × 30 cm. The gas log permeability is about 1100×10^{-3} μm^2 for the first group of cores, and about 2600×10^{-3} μm^2 for the second group. In the oil displacement experiment, a device is set on both ends of the cores to eliminate "tip effect," and 30 cm-long cores are used to ensure that the devices work effectively. In the oil displacement process, oil displacement is well done with the composite system slug as far as possible, i.e., inject combination flooding slug continuously until the produced fluid does not contain oil, and this lasts for a sufficient period of time without subsequent water driving process.

FIG. 1.2 Core section after completion of four groups of experiments.

Fig. 1.2 gives section pictures of four 2D section cores after completion of oil displacement experiments, and cores are arranged from top to bottom and the capillary number increases gradually during oil displacement. It can be clearly seen that oil displacement effect of core W1-1 is poor, core 1-5-3 better, and W1-6 the best. Sudden fingering of displacement fluid occurs along the high-permeability layer in core W1-3, and displacement effect is very good in the high-permeability layer while plenty of remaining oil remains in the medium-to-low-permeability layers.

The results of oil displacement experiments with 1D core are given in Table 1.2. It can be seen from experiments for core group 1 that residual oil saturation decreases with the increase of capillary number in oil displacement experiments. A relatively low residual oil saturation of 20.4% occurs when capillary number is about 0.108. As the capillary number continues to increase, the residual oil saturation increases and then decreases. Although there are less experiments for core group 2, the same change law appears. The permeability of core group 2 is relatively high, and the lowest residual oil saturation obtained in experiment is relatively low.

The experiment results show that the corresponding change law between residual oil saturation and capillary number is correct, which is obtained through simulation calculation under the high capillary number condition, and the classical capillary number experimental curve does not provide the description of change characteristics of residual oil saturation after limit capillary number is exceeded, which is its defect.

1.3 Calculation and study on capillary number curve shape under high capillary number condition during displacement

As numerical simulation study reveals that "two different driving conditions exist in the combination flooding process," which is further verified through oil displacement experiments, it is certain that UTCHEM software describes

TABLE 1.2 Results for oil displacement experiments in 1D model.

Core group	Experiment no.	Interfacial tension (mN/m)	System viscosity (mPa s)	Injection speed (mL/min)	Seepage velocity (m/s)	Recovery percent of reserves (%)	Capillary number N_c	Residual oil saturation (%)
1	1.1	1.29×10^{-2}	15.8	0.6	2.03×10^{-5}	66.6	2.48×10^{-2}	25.3
	1.2	1.29×10^{-2}	22	0.6	1.89×10^{-5}	68.1	3.22×10^{-2}	22.8
	1.3	1.29×10^{-2}	20	1.2	3.81×10^{-5}	72.4	5.90×10^{-2}	21.1
	1.4	2.48×10^{-3}	12.8	0.6	2.10×10^{-5}	73.0	1.08×10^{-1}	20.4
	1.5	2.48×10^{-3}	16.2	0.6	1.92×10^{-5}	64.9	1.26×10^{-1}	26.4
	1.6	2.48×10^{-3}	21.9	0.6	1.9×10^{-5}	69.5	1.68×10^{-1}	22.7
	1.7	2.48×10^{-3}	16.1	1.2	3.82×10^{-5}	71.1	2.48×10^{-1}	21.9
	1.8	2.48×10^{-3}	20	1.2	3.87×10^{-5}	70.9	3.12×10^{-1}	21.9
2	1.9	1.29×10^{-2}	22.3	0.6	1.54×10^{-5}	75	2.66×10^{-2}	19.6
	1.10	1.29×10^{-2}	21.7	1.2	3.07×10^{-5}	77.4	5.17×10^{-2}	17.5
	1.11	2.48×10^{-3}	11.9	0.6	1.49×10^{-5}	68	7.17×10^{-2}	24.7
	1.12	2.48×10^{-3}	16.4	0.6	1.52×10^{-5}	73.8	1.00×10^{-1}	20.4
	1.13	2.48×10^{-3}	22.2	0.6	1.56×10^{-5}	70.6	1.40×10^{-1}	23.2

capillary number curve in a relatively correct way. Enlightened by this, it is necessary to use a numerical simulation method to further study the capillary number curve to have an understanding of change law of capillary number curve shape under high capillary number condition. This is beneficial to provide guides for capillary number oil displacement experiments and study and analysis on change law of combination flooding residual oil saturation.

Generally, capillary number experiments are required to be done with 1D "homogeneous" cores, and structural analysis should be performed for 1D "homogeneous" cores prior to calculation and study. Cores contain pores with different pore throat radii and a certain amount of micro pore paths where residual oil saturation is relatively high even at low-tension system displacement condition. In simulation calculation, the 1D core model should be regarded as parallel capillary tube combination model with three pore radii including "large, medium and small radii" for calculation and study. In fact, capillary tubes with three pore radii represent average permeability of micro pore channels in three grades. As pores with different grades have different percentages, general permeability of cores is different. The smaller the percentage of low permeability is, the higher the general average permeability is. Under the guide of this design idea, three cores with different general permeability are designed. Table 1.3 gives their detailed data in three grades.

The model used in calculation and study is $9 \times 1 \times 3$ grid, $\Delta X = \Delta Y = 22.098$ m, and the thickness proportion for three layers in Z direction is determined as the thickness percentage of corresponding permeability grade of each model, as shown in Table 1.3. As the change of residual oil after ternary system driving is to be studied here, the oil displacement process is designed to be: combination flooding follows water flooding at 0.6 Vp, with a total fluid volume injected of 14.42 Vp. No oil production is required at the end of schemes.

Fig. 1.3 shows the relation curve between capillary number and residual oil saturation under high capillary number condition based on calculation results of model *A*.

TABLE 1.3 Detailed data for cores with different permeability in three grades.

	Average permeability in three grades and thickness percentage of corresponding layers (%)			Average permeability
Model	$1 \times 10^{-3}\,\mu m^2$	$100 \times 10^{-3}\,\mu m^2$	$1000 \times 10^{-3}\,\mu m^2$	$1 \times 10^{-3}\,\mu m^2$
A	33.33	33.33	33.33	367
B	25.00	37.50	37.50	413
C	16.67	41.67	41.67	459

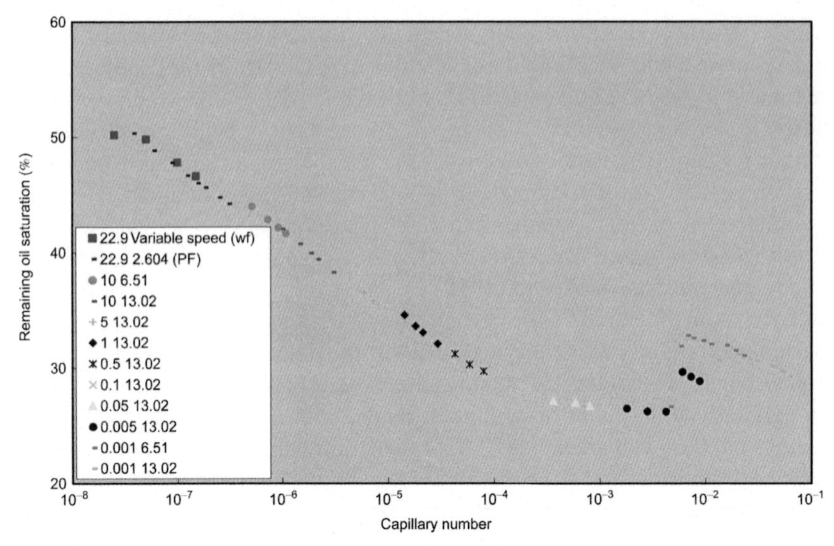

FIG. 1.3 Relation curve between capillary number and residual oil saturation for model *A*.

The numbers in the first column in the figure are interfacial tension of the system in mN/m, and those in the second column are fluid injection speed in m^3/d.

The general shape of capillary number curve can be seen in the figure. Simulation calculation starts with variable speed water flooding, and the first several calculation points are in level; under high-speed water flooding and polymer flooding conditions, changes occur in residual oil saturation, and we obtain limit capillary number N_{cc}. Starting from this point, the residual oil saturation shows apparent decrease with the increase of capillary number, and a new capillary number limit appears at capillary number value of 0.001. Starting from this point, the residual oil saturation does not decrease and remains flat with the increase of capillary number. Such change continues until the limit capillary number N_{ct}. After that, residual oil saturation shows complex changes with the increase of capillary number. It can be seen clearly from the figure that one capillary number corresponds to multiple residual oil saturation values, that is, capillary number corresponds to multiple residual oil saturation values as capillary number increases. These values are distributed in a regular way. For experiments with the same interfacial tension, fluid injection speed, and different system viscosity, the residual oil saturation values are on one curve, and as system viscosity and capillary number increase, the residual oil saturation increases to a certain maximum value and then decreases. Under the condition of the same interfacial tension and different fluid injection speed, the residual oil saturation curve shows nearly in parallel, with high speed ones located in the lower part; as the magnitude of interfacial tension decreases, the curves move to the right in parallel.

It is noted that capillary number curve shows a flat and straight section and the limit capillary number N_{ct} corresponds to the right end point of the flat and straight section, and the left end point corresponds to a new limit capillary number. For the purpose of description, the original capillary number limit point N_{ct} is called N_{ct1} and the other capillary number limit point N_{ct2}. Low residual oil saturation values are between the two limit values, which are called residual oil saturation S_{or}^{H} corresponding to limit capillary number N_{ct1} (or N_{ct2}). The calculation result here is about 26.11%.

Calculation and study are also done for models *B* and *C*. Three curves show the same shape, and the corresponding residual oil saturation values S_{or}^{H} on the three curves are 26.11%, 19.66%, and 13.10% respectively, which corresponds to the average permeability of three core groups that rise in turn, and matches with two 1D core oil displacement experiment results with different permeability given in the previous section.

1.4 Re-study on experimental curve of relations between capillary number and residual oil saturation

1.4.1 Measurement of experimental curve of relations between capillary number and residual oil saturation

Artificial cores in the 1D model are used in the oil displacement experiment. The requirements for cores are as follows: each core is homogeneous, and different cores are comparable; a device must be set on both ends of cores to eliminate "tip effect"; core size is 4 cm × 4 cm × 30 cm. Substituting Darcy's formula in the capillary number definition formula, we obtain:

$$N_c = \frac{V \cdot \mu_w}{\sigma_{ow}} = \left[\frac{KK_{rw}}{\mu_w} \cdot \frac{\Delta p}{L}\right] \frac{\mu_w}{\sigma_{ow}} = \frac{KK_{rw}}{\sigma_{ow}} \cdot \frac{\Delta p}{L}. \tag{1.7}$$

It can be seen from the formula that when core length is relatively big, experiments can be well controlled, which is beneficial to the precision of calculated capillary number values.

It is confirmed from oil displacement experiment that the residual oil saturation of combination flooding has no relation to water flooding before combination flooding on the cores, which are saturated by water and oil and not flooded by chemical combination system before. So, in the oil displacement experiment, water flooding begins first until a water cut of 90%, followed by combination flooding, then the combination flooding slug is injected until produced fluid does not contain oil. This lasts for a period of time before the experiment is terminated and there is no subsequent water flooding. Experiments were done under oil-water condition at Daqing Oilfield. As the capillary number range is large in the oil displacement experiment, water flooding and combination flooding with multiple combinations of oil displacement systems are used. The combination flooding system includes a ternary system with alkaline, but a binary system without alkaline is used more frequently.

It should be noted that seepage velocity V in the calculation capillary number value is expressed below:

$$V = \frac{Q}{A \times (1 - S_{or})}. \tag{1.8}$$

Table 1.4 gives basic experimental parameters and results. Fig. 1.4 shows the curve of relation between capillary number and residual oil under high capillary number condition based on the experiment results. To differentiate from previous experiment curves, the new curve is named "experimental curve QL of relation between capillary number and residual oil saturation."

The capillary number value at the left end point of curve section "I" in Fig. 1.4 is 1.09×10^{-6}, and the corresponding residual oil saturation is about 28.7%. As the capillary number increases continuously, the residual oil saturation decreases slightly. When the capillary number is around 1.15×10^{-4}, the residual oil saturation is about 27.3%. This process occurs generally under water flooding condition; combination flooding starts at this moment, and as the capillary number increases further, the residual oil saturation decreases dramatically. When the capillary number is around 2.7×10^{-3}, the corresponding residual oil saturation value is about 18.5%. When the capillary number increases further until the right end point of curve section "I" at which the capillary number value is about 7.15×10^{-2}, the residual oil saturation is in a stable state, with nearly no increase or decrease. The driving process that corresponds to points on curve section "I," is called a "Type I" driving condition, and the residual oil saturation value is a "single value" function of capillary number. The conversion point on the curve at where water flooding is converted to combination flooding, the turning point in combination flooding process at which the residual oil saturation decreases and then becomes stable, and the right end point of curve "I" are called "water flooding critical capillary number N_{cc}," "limit capillary number N_{ct2}," and "limit capillary number N_{ct1}."

The capillary number continues to increase from N_{ct1} and the oil displacement process is converted to a "Type II" driving condition. It can be seen from experiment results that multiple residual oil saturation values correspond to the same capillary number value. Curve II_1 shows that when the capillary number begins to increase, the residual oil saturation value increases suddenly. After the residual oil saturation value reaches the maximum value, it decreases as the capillary number increases; curve II_2 (fluid injection speed 0.9 mL/min, interfacial tension 6.8×10^{-3} mN/m), curve II_3 (fluid injection speed 0.6 mL/min, interfacial tension 1.5×10^{-3} mN/m), and curve II_4 (fluid injection speed 0.9 mL/min, interfacial tension 1.5×10^{-3} mN/m) are all obtained under the condition of fixed system interfacial tension, fixed fluid injection speed, and variable system viscosity. The oil displacement system interfacial tensions of curves II_3 in the oil displacement experiment are close to that of II_4, and the fluid injection speed

TABLE 1.4 Basic data for oil displacement experiment of capillary number.

Core no.	Injection speed (mL/min)	System viscosity (mPa s)	Interfacial tension (mN/m)	Seepage velocity (10^{-5} m/s)	Recovery percent of reserves by water flooding (%)	Final recovery factor (%)	Capillary number	Residual oil saturation (%)	Curve marks
3-9	1.00	0.60	2.25×10^{1}	2.918	60.5	60.50	1.09×10^{-6}	28.6	I
3-16	1.00	0.60	2.25×10^{1}	3.180	63.1	63.10	1.19×10^{-6}	28.7	
3-94	0.70	5.30	1.50×10^{0}	3.010	44.4	64.32	1.06×10^{-4}	28.4	
3-8	0.60	6.40	1.20×10^{0}	2.350	41.6	67.31	1.25×10^{-4}	26.2	
3-10	0.60	3.56	1.11×10^{-1}	2.150	37.2	70.29	6.89×10^{-4}	23.3	
3-73	0.60	3.56	1.11×10^{-1}	2.200	37.2	68.50	7.04×10^{-4}	24.1	
3-84	0.60	20.40	1.70×10^{-1}	2.150	45.6	75.00	2.58×10^{-3}	19.2	
3-89	0.60	20.40	1.70×10^{-1}	2.340	46.1	77.61	2.81×10^{-3}	19.9	
3-51	0.80	23.60	8.80×10^{-2}	2.800	0[a]	76.52	7.51×10^{-3}	18.3	
3-63	0.80	23.60	6.50×10^{-2}	2.920	0[a]	76.54	7.83×10^{-3}	18.7	
3-35	0.80	17.90	5.83×10^{-2}	2.700	46.3	74.78	8.29×10^{-3}	19.1	
3-61	0.70	24.00	6.50×10^{-2}	2.330	46.3	75.40	8.59×10^{-3}	17.8	
3-29	0.60	11.90	6.40×10^{-3}	2.050	56.7	76.41	3.81×10^{-2}	18.3	
3-70	1.00	11.50	9.80×10^{-3}	3.650	51.1	75.91	4.28×10^{-2}	19.3	

Continued

TABLE 1.4 Basic data for oil displacement experiment of capillary number—cont'd

Core no.	Injection speed (mL/min)	System viscosity (mPa s)	Interfacial tension (mN/m)	Seepage velocity (10^{-5} m/s)	Recovery percent of reserves by water flooding (%)	Final recovery factor (%)	Capillary number	Residual oil saturation (%)	Curve marks
3-83	1.00	11.50	9.80×10^{-3}	3.640	50.8	75.70	4.27×10^{-2}	19.4	
3-30	0.70	24.10	1.00×10^{-3}	2.500	48.1	75.00	6.03×10^{-2}	19.5	
3-27	0.55	26.40	7.20×10^{-3}	1.940	47.9	75.81	7.12×10^{-2}	18.6	
3-44	0.60	24.00	7.20×10^{-3}	2.160	44.5	74.00	7.20×10^{-2}	20.1	II_1
3-45	0.60	24.00	7.20×10^{-3}	2.170	44.3	73.43	7.25×10^{-2}	21.1	
3-18	0.60	24.30	7.20×10^{-3}	2.270	52.6	74.02	7.65×10^{-2}	20.5	
3-53	0.60	29.20	6.80×10^{-3}	2.327	44.6	70.24	9.99×10^{-2}	22.3	
3-11	0.80	16.50	5.00×10^{-3}	2.860	46.3	69.50	9.43×10^{-2}	24.0	
3-14	0.55	26.40	7.20×10^{-3}	2.250	45.6	64.38	8.23×10^{-2}	27.8	
3-15	0.80	16.50	5.00×10^{-3}	3.080	37.2	66.25	1.02×10^{-1}	27.9	
3-72	0.90	19.50	6.80×10^{-3}	3.290	44.4	71.94	9.44×10^{-2}	20.8	II_2
3-23	0.90	28.60	6.80×10^{-3}	3.391	47.1	68.20	1.43×10^{-1}	23.6	
3-85	0.90	38.20	6.80×10^{-3}	3.495	44.4	66.50	1.96×10^{-1}	25.0	
3-54	0.90	38.30	6.80×10^{-3}	3.566	44.4	65.38	2.01×10^{-1}	26.9	

3-17	0.90	53.50	6.80×10^{-3}	3.513	35.7	70.32	2.76×10^{-1}	22.9	
3-62	0.90	53.50	6.80×10^{-3}	3.388	42.1	70.48	2.67×10^{-1}	22.2	
3-5	0.60	10.90	1.50×10^{-3}	2.420	44.8	67.84	1.76×10^{-1}	25.1	II_3
3-69	0.60	10.90	1.50×10^{-3}	2.350	45.7	67.48	1.71×10^{-1}	25.7	
3-81	0.60	15.10	1.50×10^{-3}	2.160	42.3	73.12	2.17×10^{-1}	20.0	
3-31	0.60	15.10	1.50×10^{-3}	2.210	41.2	73.64	2.22×10^{-1}	20.6	
3-3	0.60	21.90	1.70×10^{-3}	2.260	41.0	68.40	2.91×10^{-1}	23.7	
3-48	0.60	21.90	1.70×1010^{-3}	2.240	44.5	68.77	2.89×10^{-1}	23.9	
3-1	0.60	37.40	1.70×1010^{-3}	2.190	49.3	74.49	4.81×10^{-1}	19.6	
3-26	0.60	37.40	1.70×1010^{-3}	2.080	50.3	78.71	4.57×10^{-1}	15.9	
3-50	0.90	7.20	1.50×1010^{-3}	3.220	40.0	71.27	1.55×10^{-1}	22.2	II_4
3-21	0.90	11.10	1.50×1010^{-3}	3.580	53.3	69.67	2.65×10^{-1}	23.0	
3-82	0.90	17.30	1.50×1010^{-3}	3.230	40.1	71.49	3.72×10^{-1}	22.1	
3-57	0.90	17.30	1.50×1010^{-3}	2.460	40.4	68.59	2.83×10^{-1}	25.2	
3-68	0.90	24.10	1.50×1010^{-3}	2.990	46.8	77.90	4.80×10^{-1}	17.3	

[a] *The data is obtained from experiments of relative permeability curve test, with no water flooding.*

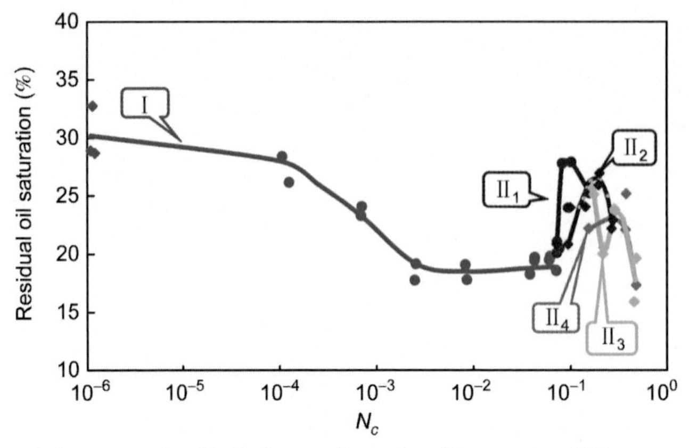

FIG. 1.4 Relation curve of residual oil saturation and capillary number (QL).

of the latter in the experiment is higher than that of the former. Two curves are close in shape, and the curve for the latter is below the curve for the former, which shows the effect of fluid injection speed; the fluid injection speeds in experiment for curves II_2 and II_4 are the same. The system interfacial tension for curve II_2 is higher and curve II_2 is to the left of curve II_4, which shows the effect of interfacial tension change. It can be inferred that similar curves can be obtained by changing conditions, and such curves are numerous and distributed in a certain area.

1.4.2 Re-verification experiment for key change points on capillary number experiment curve

Comparison of capillary number curve QL with the classical capillary number curve shows distinct differences. The main difference is that there is a flat and straight section on curve QL between N_{ct2} and N_{ct1}, which corresponds to the same residual oil saturation value S_{or}^{H}. Residual oil increases when the capillary number is higher than capillary number limit point N_{ct1}. In order to reproduce this experiment result, another two groups of cores are used to complete oil displacement experiments to verify correctness and to assess impact factors for key parameters of experiment curves. Experiments are done under oil-water condition at Daqing Oilfield. One group of cores is 1D homogeneous core, with an average water phase permeability of $460 \times 10^{-3}\ \mu m^2$. To assess the impacts of core shape on experiment results, cylindrical cores are used. Generally, capillary number curve experiments are done with a 1D homogeneous core. To assess the impacts of heterogeneous section at the same time, a group of square cores with heterogeneous section and small interlayer permeability grade difference is used in experiments, and the permeability of three layers is 200, 600, and $1200 \times 10^{-3}\ \mu m^2$, with average water phase permeability of

784×10^{-3} μm^2 measured in experiment. Five pairs of parallel and efficient oil displacement experiments are completed in each group. Table 1.5 gives experiment results.

It can be clearly seen from Table 1.5 that capillary number limit points N_{ct2} and N_{ct1} can be measured in two groups of experiments. When the capillary number is less than the limit capillary number N_{ct2}, it corresponds to relatively high residual oil saturation S_{or}. Under the condition that the capillary number is higher than limit capillary number N_{ct1}, relatively high residual oil saturations are obtained in experiments. To further understand the differences in the results of experiments with different permeability cores, Table 1.6 summarizes values of the limit capillary numbers N_{ct2} and N_{ct1} and residual oil saturation S_{or} measured in three groups of experiments.

Cylindrical core experiments do not show peculiar changes, eliminating the doubt about square core experiment results. It is pleasing that the results of experiments with heterogeneous section cores are close to the experiment results with homogeneous cores. With careful observation, the data of average permeability of core group 2# is closer to those of core group 1#. However, data of three items are close to those of core group 3#; this is obvious that the core heterogeneous is aggravated because core group 2# is heterogeneous section core.

Verification experiments verify that capillary number curves of core experiments with different permeability have basically the same shape, thus it can be believed that capillary number experiment curve QL reflects correctly the relations between capillary number and residual oil saturation in oil displacement experiments.

1.4.3 Further study on high residual oil saturation occurring under high capillary number condition

Fig. 1.5 shows the corresponding relation of nonwetting phase saturation and pore throat radius measured by mercury porosimetry with cores used in experiments of capillary number curves QL.

Table 1.7 shows the core experiment data with similar pore volume from Section 1.4.1. The capillary number of the first seven experiments in Table 1.7 is between limit capillary number N_{ct1} and N_{ct2}. The cumulative oil production is close with each other, with an average value of 99.31 mL. During combination flooding under high capillary number condition, the oil resisted by capillary force is activated and then flooded away. Generally speaking, the larger the capillary number, the smaller pore throat radius in which the oil can be moved. According to the analysis in Fig. 1.5, if V represents pore space that core pore throat radius is higher than 7.0 μm, then the volume of V is about 100 mL. Apparently, in this combination flooding experiment, the oil in the V pore space is driven. The capillary number of other experiments is higher than the limit capillary number N_{ct1}, the lowest value of the residual oil saturation is

TABLE 1.5 Basic data for oil displacement experiment of capillary number.

Core no.	Gas log permeability (10^{-3} μm^2)	Injection speed (mL/min)	System viscosity (mPa s)	Interfacial tension (mN/m)	Seepage velocity (m/s)	Recovery percent of reserves by water flooding (%)	Final oil recovery factor (%)	Capillary number (N_c)	Residual oil saturation (%)
Y19-49	1315	0.7	11.8	1.46	7.29×10^{-5}	41.03	52.4	5.89×10^{-4}	37.31
Y19-87	1331	0.7	11.8	1.46	7.02×10^{-5}	37.41	52.2	5.67×10^{-4}	34.88
Y19-25	1213	0.5	6.0	1.95×10^{-1}	4.35×10^{-5}	40.98	60.0	1.34×10^{-3}	30.50
Y19-8	1269	0.5	6.0	1.95×10^{-1}	4.54×10^{-5}	41.80	59.05	1.40×10^{-3}	32.66
Y19-5	1245	0.5	16.0	6.80×10^{-2}	4.3×10^{-5}	37.88	60.91	1.01×10^{-2}	31.46
Y19-64	1294	0.5	16.0	6.80×10^{-2}	4.28×10^{-5}	40.50	57.54	1.01×10^{-2}	32.86
Y19-78	1284	0.5	22.3	6.80×10^{-3}	4.61×10^{-5}	43.28	59.02	1.51×10^{-1}	31.93
Y19-3	1251	0.5	22.0	6.20×10^{-3}	4.27×10^{-5}	36.77	60.65	1.51×10^{-1}	30.12

Y19-18	1232	0.7	32.5	1.04×10^{-2}	6.82×10^{-5}	33.67	51.0	2.13×10^{-1}	37.22
Y19-60	1306	0.7	22.3	6.80×10^{-3}	7.08×10^{-5}	30.49	52.13	2.32×10^{-1}	37.92
S2-25	2357	0.9	21.8	1.60	4.45×10^{-5}	25.37	53.53	6.06×10^{-4}	35.10
S2-28	2345	0.9	21.8	1.60	4.34×10^{-5}	24.11	54.38	5.91×10^{-4}	34.76
S2-23	2369	0.9	10.0	1.86×10^{-1}	3.88×10^{-5}	26.46	66.50	2.09×10^{-3}	23.62
S2-29	2342	0.9	10.0	1.86×10^{-1}	3.90×10^{-5}	27.45	65.69	2.10×10^{-3}	24.62
S2-12	2358	0.9	10.0	1.15×10^{-1}	3.82×10^{-5}	25.47	66.32	1.02×10^{-2}	24.29
S2-16	2355	0.9	10.0	1.15×10^{-1}	3.52×10^{-5}	27.50	66.40	9.40×10^{-3}	24.00
S2-20	2375	0.6	18.6	5.50×10^{-3}	3.38×10^{-5}	26.41	69.25	1.11×10^{-1}	24.14
S2-22	2376	0.8	5.00	2.00×10^{-3}	4.35×10^{-5}	25.29	60.47	1.04×10^{-1}	24.78
S2-27	2375	0.7	4.60	9.60×10^{-4}	3.29×10^{-5}	47.60	53.37	1.62×10^{-1}	36.19
S2-21	2375	0.9	4.60	9.60×10^{-4}	4.18×10^{-5}	36.91	55.09	2.08×10^{-1}	35.80

TABLE 1.6 Summary of key data for capillary number curve oil displacement experiments of different core groups.

Core group no.	Average K ($10^{-3}\mu m^2$)		N_{ct1}		N_{ct2}		S_{or} (%)
	Gas logging	*Water logging*	*Upper limit*	*Lower limit*	*Upper limit*	*Lower limit*	
1	2839	871	7.12×10^{-2}	7.20×10^{-2}	7.04×10^{-4}	2.58×10^{-3}	18.67
2	2375	784	1.11×10^{-1}	1.62×10^{-1}	5.91×10^{-4}	2.09×10^{-3}	24.24
3	1274	460	1.51×10^{-1}	2.13×10^{-1}	5.89×10^{-4}	1.34×10^{-3}	31.59

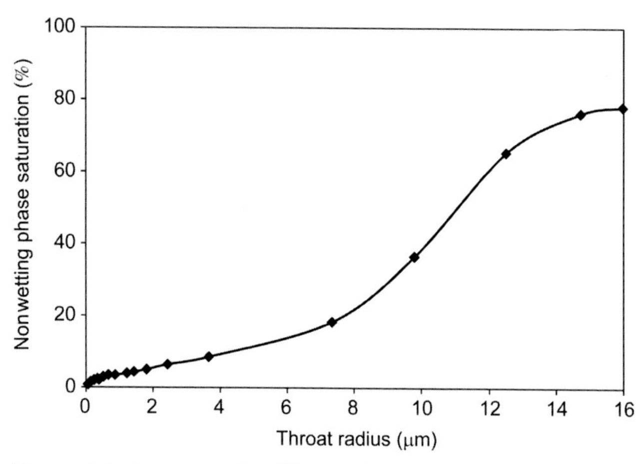

FIG. 1.5 Mercury injection curves of capillary number curve displacement experiment (QL).

22.2% when oil displacement experiment process ends, and the oil production is obviously lower. Analysis of the experiment data of core 3–53 shows that the capillary number is 0.0999, the oil in the *V* pore space should be flooded, and the oil production should be at least 100 mL, while the actual is only 86.4 mL. It is 12.91 mL lower than the average value of 99.31 mL of the first seven experiments. Analysis of the water production shows the total volume of fluid injected is 557.6 mL; if the oil in the *V* pore space is flooded out, then there should be 99.31 mL of displacement fluid detained in the *V* pore space and fluid produced should be 458.29 mL, so the remaining fluid produced (12.91 mL) is the formation water. It is worth investigating why the oil removed from the *V* space has not been produced, where it remains, and where the reservoir water in the produced fluid came from.

The results of other core experiments are nearly the same. The common feature is that oil production is significantly reduced. The minimum reduction is 9.91 mL, corresponding to the production of oil reservoir water of the same volume.

This concludes that under the condition of capillary number higher than N_{ct1}, the fluid has been displaced in a tiny pore throat where oil and water coexist, and wettability converses happen in a partial pore throat, which used to be water wet, and the irreducible water in the pore throat is activated and driven away, which is the produced formation water. The wettability of pore throat converses into oil wet, and the flowing crude oil is trapped as irreducible oil, which causes the oil production to decrease and the residual oil saturation to increase after flooding.

TABLE 1.7 Partial core data of capillary number curve QL experiment.

Core no.	Irreducible oil saturation (%)	Pore volume (mL)	Cumulative fluid injection (mL)	Cumulative oil production (mL)	Reduction in oil production (mL)	Cumulative water production no. 1 (mL)	Cumulative water production no. 2 (mL)	Incensement in water production (mL)
3-84	19.2	167	651.3	96.0		555.3		
3-89	19.9	169	659.1	104.0		555.1		
3-61	17.8	164	672.4	95.4		577.0		
3-70	19.3	165	924.0	100.2		823.8		
3-83	19.4	165	924.0	102.2		821.8		
3-30	19.8	168	621.6	99.6		522.0		
3-27	18.6	168	520.8	97.8		423.0		
3-53	22.3	164	557.6	86.4	12.91	471.2	458.29	12.91
3-14	27.8	164	508.4	82.4	16.91	426.0	409.09	16.91
3-23	23.6	169	811.2	89.0	10.31	722.2	711.89	10.31
3-85	25.0	168	823.2	85.0	14.31	738.2	723.89	14.31
3-54	26.9	167	718.0	85.0	14.31	633.0	618.69	14.31
3-17	22.9	161	740.6	87.2	12.11	653.4	641.29	12.11
3-62	22.2	165	742.5	87.4	11.91	655.1	643.19	11.91
3-5	25.1	160	592.0	84.8	14.51	507.2	492.69	14.51
3-69	25.7	165	610.5	88.4	10.91	522.1	511.19	10.91
3-3	23.7	168	621.6	86.2	13.11	535.4	522.29	13.11
3-48	23.9	170	629.0	89.4	9.91	539.6	529.69	9.91

On further analysis, this wettability conversion phenomenon has begun to appear when the capillary number is higher than N_{ct2}. While the capillary number is between N_{ct1} and N_{ct2}, the volume of newly started bound oil due to the capillary number increasing is equal to the volume space of wettability conversion to oil wet, that is to say, the new release of crude oil is the same as the amount of entrapped space with wettability conversion. When flooding process is terminated, the residual oil saturation does not increase or decrease, and capillary curves show a “flat segment”; when the capillary number is higher than N_{ct1}, the volume of newly started bound oil due to the capillary number increasing is lower than the volume space of wettability conversion to oil wet, and the incremental of released oil is less than the entrapped oil, appear to the residual oil saturation increase as the capillary number increasing. Along with decreasing of the interfacial tension σ_{ow} between oil and water, the pore throat radius of activated bound oil-water reduces. Under certain conditions of interfacial tension, the pore throat radius of activated bound oil-water is fixed, and the amount of oil bound activated is fixed. For a portion of the space releasing bound water converted into oil-wet space, the flowing crude oil from the flowable oil spaces would be captured and converted into the bound oil, and the amount of flowable oil determined the amount of oil to be captured. The amount of flowable oil depends on the flow velocity V of the drive phase (aqueous phase). The higher the flow speed, the more the amount of oil driven away. The oil remaining in flowable space reduces, the amount of captured oil is relatively small, and the corresponding residual oil saturation is relatively low. At the same flow rate, water phase viscosity μ_w adjusts the flow ratio, changes the amount of oil flowing away, and affects the amount of crude oil being captured. Different viscosity of flooding system can form different residual oil saturation to obtain capillary numbers curve with the condition of given interfacial tension and flow velocity. That is why the capillary number curve forks and changes regularly.

1.5 Re-study of relative permeability curve in combination flooding

As the experimental curve QL is obtained, we have a clear understanding of the relation between capillary number and residual oil. However, it is not enough to understand only the change of residual oil saturation. We should also further understand the change of relative permeability curve in combination flooding process. Many references[7–12] describe the issue of relative permeability curve of combination flooding. Based on the study of capillary number experimental curve QL, combination flooding relative permeability curve is studied through experimentation to provide mathematic description of relative permeability curve and to determine the key parameters in the expression.

1.5.1 Expression determination of relative permeability curve in combination flooding

Based on the experimental curve QL, with reference to the mathematic description of relative permeability curve in the UTCHEM software, a new expression of combination flooding relative permeability curve is worked out, corroborated by combination flooding experiments and simulated calculation study.

When water saturation S_w meets $S_{wr}^L \le S_w \le 1 - S_{or}$, the "normalized" water saturation S_{nw} is calculated through the expression below:

$$S_{nw} = \frac{S_w - S_{wr}^L}{1 - S_{wr}^L - S_{or}}. \tag{1.9}$$

The relative permeability curves for water and oil phases can be expressed respectively:

$$K_{rw} = K_{rw}^0 (S_{nw})^{n_w}, \tag{1.10}$$

$$K_{ro} = K_{ro}^0 (1 - S_{nw})^{n_o}. \tag{1.11}$$

Corresponding to different change range of capillary number N_c, the values of key parameters are as follows.

(1) Residual oil saturation S_{or}.

$$S_{or} = \begin{cases} S_{or}^L, & N_c \le N_{cc}, \\ S_{or}^L - \dfrac{N_c - N_{cc}}{N_{ct2} - N_{cc}} \left(S_{or}^L - S_{or}^H\right), & N_{cc} \le N_c \le N_{ct2}, \\ S_{or}^H, & N_{ct2} \le N_c \le N_{ct1}, \\ S_{or}^L / (1 + T_o \times N_c), & N_c > N_{ct1}. \end{cases} \tag{1.12}$$

(2) End point value K_{rl}^o $l = w, o$.

$$k_{rl}^0 = \begin{cases} K_{rl}^L, & N_c \le N_{cc}, \\ K_{rl}^L + \dfrac{N_c - N_{cc}}{N_{ct2} - N_{cc}} \left(K_{rl}^H - K_{rl}^L\right), & N_{cc} \le N_c \le N_{ct2}, \\ K_{rl}^H, & N_c \ge N_{ct2}. \end{cases} \tag{1.13}$$

(3) Exponent value n_l $l = w, o$.

$$n_l = \begin{cases} n_l^L, & N_c \le N_{cc}, \\ n_l^L + \dfrac{N_c - N_{cc}}{N_{ct2} - N_{cc}} \left(n_l^H - n_l^L\right), & N_{cc} \le N_c \le N_{ct2}, \\ n_l^H, & N_c \ge N_{ct2}. \end{cases} \tag{1.14}$$

Comparison of the above mathematic expression of combination flooding relative permeability curve with the expression given in UTCHEM shows that it has the former's advantages, but it also has important differences. Based on capillary number experimental curve QL, the relation between capillary number and residual oil saturation under water phase displacement condition is given, and then corresponding relative permeability curve is given. This treatment is reasonable and can simplify calculations greatly.

1.5.2 Determination of relative permeability curve parameters

With determination of the expression of the relative permeability curve, the experimental measure method for relative permeability curve parameters is also determined. The limit capillary numbers N_{cc}, N_{ct1}, and N_{ct2} can be obtained from "capillary number curve QL," and other parameters can be determined through oil displacement experiments. With cores used in the "capillary number curve QL" experiment, the water flooding relative permeability curve and combination flooding relative permeability curve are determined via the nonstable relative permeability curve determination method at the condition of oil reservoir temperature and ambient pressure. Relatively high fluid injection speed should be used in water flooding experiments so that the capillary number in the oil displacement process is as close as possible to the limit capillary number N_{cc}. The experimental data of the water flooding relative permeability curve are shown in Table 1.4 and respond to core numbers 3-9 and 3-16. The capillary number in the oil displacement process of combination flooding should be strictly controlled between limit capillary numbers N_{ct2} and N_{ct1}. The experimental data of the combination flooding relative permeability curve shown in Table 1.4 responds to core number 3-51 and 3-63, with no water flooding in the prior period. The determined relative permeability curve data should be analyzed using the regression method to obtain end point and exponent values.

Table 1.8 gives data for seven groups of relative permeability curves, among which $N_{cc}=0.0001$, $N_{ct1}=0.0712$, and $N_{ct2}=0.0025$ are obtained through "capillary number curve QL." In the seven groups of parameters, Groups 1 and 4 are determined through experiment, Groups 2 and 3 are obtained through interpolation calculation, and the last three groups are obtained after determining parameter T_o value in combination of experiment results and calculation study. Fig. 1.6 shows relative permeability curves of the first four groups. The curves of the last three groups are not shown because they differ from the curves of the first four groups only in residual oil saturation S_{or}.

To correspond to "capillary number curve QL," the associated relative permeability curve is named "combination flooding relative permeability curve QL."

1.5.3 Two ways to describe relative permeability curves

The previously given description of "combination flooding relative permeability curve QL" is based on the hypothesis that "irreducible water saturation is constant."

TABLE 1.8 Parameters for relative permeability curves.

N_c		S_{wr}^L	S_{or}	K_{ro}	n_o	K_{rw}	n_w
$N_c \le N_{cc}$		0.24	0.285	1	1.95	0.255	3.75
$N_c = 0.001$		0.24	0.2475	1	1.92	0.534	4.35
$N_c = 0.00175$		0.24	0.21625	1	1.895	0.767	4.85
$N_{ct2} \le N_c \le N_{ct1}$		0.24	0.185	1	1.87	1	5.35
$T_o = 1.2$	$N_c = 0.1$	0.24	0.254	1	1.87	1	5.35
	$N_c = 0.4$	0.24	0.192	1	1.87	1	5.35
	$N_c = 0.5$	0.24	0.178	1	1.87	1	5.35

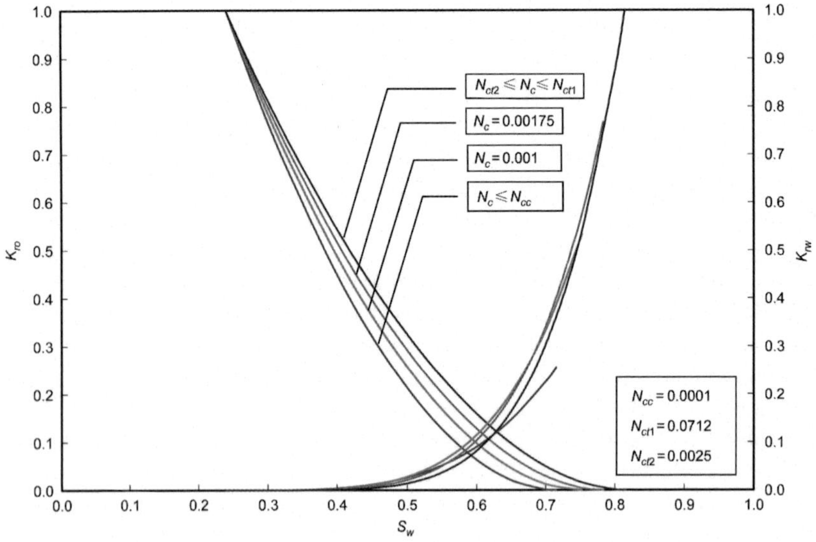

FIG. 1.6 Combination flooding relative permeability curves.

After the study in Section 1.4.3, we describe the relative permeability curves in another way; in a range of capillary number, we describe irreducible water saturation dynamically. On the basis of the study above, we get the following:

(1) Under the condition that capillary number is higher than N_{ct1} and the corresponding irreducible oil saturation S_{or} is higher than S_{or}^H, we should revise the irreducible water saturation S_{wr}^L in Eq. (1.9), in which the reduction should be the same with the incremental value of residual oil saturation S_{or} relative to S_{or}^H.

(2) We should consider the change of flooding process. If in some place, capillary number is between N_{ct1} and N_{ct2}, and it has undergone the "Type II" driving condition, which is when capillary number is higher than N_{ct1} and the corresponding residual oil saturation is higher than S_{or}^{H}, then the irreducible water saturation S_{wr}^{L} and residual oil saturation S_{or} should also be changed.

1.6 Combination flooding software IMCFS

A combination flooding software named Improved Mechanism of Combination Flooding Simulation (IMCFS) is developed based on relative permeability curve QL and corresponding capillary number curve QL. As oil layer property varies in different zones or different intervals, different capillary number curves and different relative permeability curves should be used. Several "pools" are set up in this software to describe different oil layers, and each pool has its own capillary number curve data and relative permeability curve data.

The software has been successfully applied in combination flooding field test studies and laboratory experiments, and some important achievements have been obtained. Subsequent chapters will introduce the main achievements.

1.7 Conclusions

(1) The concept proposal of the capillary number lays a theoretical foundation for the application and study of enhanced oil recovery technology by combination flooding. The obtainment of capillary number curves and the development of chemical flooding numerical simulation software UTCHEM pushes forward the study and application of combination flooding technology.
(2) Experimental study and field test analysis indicate that the relation between residual oil saturation and capillary number is complicated under high capillary number condition. The classical capillary number experimental curve does not describe the complicated relation between capillary number and residual oil saturation under high capillary number. This study has corrected the defect and improved it.
(3) Complete description of the curve shape showing the relation between capillary number and residual oil saturation is obtained through simulation calculation and study, providing an important reference basis for the obtainment of experiment capillary number curve.
(4) With the guide of capillary number curve obtained through calculation, a capillary number experiment curve is obtained, which is named "capillary number experiment curve QL." It fully corrects the defect in the description of classical capillary number experiment curve under high capillary number.
(5) Under the condition that capillary number is higher than N_{ct1} in the oil displacement process, wettability converse appears in the oil-water zone

where pore radius is relatively small. Partial water-wet pores became oil-wet pores. The irreducible water is flooded away, capturing flowable oil as irreducible oil. Therefore residual oil increased.

(6) Based on the capillary number experimental curve QL, with reference to the mathematic description of relative permeability curve in UTCHEM, the expression of combination flooding relative permeability curve is worked out, cooperated with combination flooding experiments and simulated calculation study. The key parameters are measured through oil displacement experiments and the associated combination flooding relative permeability curve QL is obtained.

(7) IMCFS combination flooding software is developed based on "the capillary number experiment curve QL" and "the combination flooding relative permeability curve QL," and the application of the software will further push forward the study and application of combination flooding technology.

Symbol descriptions

l	the mark of "phase"; *w* for water phase and *o* for oil phase
S_{nw}	"normalized" saturation of phase l
T_o	wettability converse affection parameter, gotten from experiment or matching
k_{rl}^0	relative permeability end point value of phase l
K_{rl}^L	relative permeability curve end point value of phase l at capillary number $N_c \leq N_{cc}$
K_{rl}^H	relative permeability curve end point value corresponding to capillary number lies between limit capillary number N_{ct2} and N_{ct1}
n_l	relative permeability curve exponent value of phase l
n_l^L	relative permeability curve exponent value of phase l at capillary numbers $N_c \leq N_{cc}$
n_l^H	relative permeability curve exponent value of phase l corresponding to capillary number lies between limit capillary number N_{ct2} and N_{ct1}
V	seepage velocity of displacement phase (m/s)
μ_w	viscosity of displacing phase (mPa s)
σ_{ow}	interfacial tension between displacing phase and displaced phase (mN/m)
K	absolute permeability of core (μm^2)
K_{rw}	relative permeability of water phase
Δp	seepage pressure difference (MPa)
L	seepage distance (m)
Q	flow rate passing through cores (cm^3/s)
A	pore area of core section (cm^2)

N_c	capillary number value, dimensionless
N_{cc}	limit capillary number when the residual oil begins to flow in water flooding
N_{ct1}	limit capillary number when conversion of driving condition occurs during combination flooding process, dimensionless
N_{ct2}	limit capillary number when the residual oil value in combination flooding does not decrease any longer under "Type I" driving condition, dimensionless
S_{or}	residual oil saturation corresponding to capillary number value N_c
S_{wr}^L	irreducible water saturation under low capillary number condition, i.e., $N_c \leq N_{cc}$
S_w	water saturation
S_{or}^L	residual oil saturation under low capillary number condition, i.e., $N_c \leq N_{cc}$
S_{or}^H	lowest residual oil saturation in combination flooding process under "Type I" driving condition, i.e., residual oil saturation corresponding to capillary number lies between limit capillary number N_{ct2} and N_{ct1}
"Type I" driving status	displacement under the condition that the capillary number is less than or equal to the limit capillary number N_{ct1}
"Type II" driving status	displacement under the condition that the capillary number is higher than the limit capillary number N_{ct1}

References

1. Moore TF, Slobod RC. The effect of viscosity and capillarity on the displacement of oil by water. *Producers Monthly*. 1956;20:20–30.
2. Taber JJ. Dynamic and static forces required to remove a discontinuous oil phase from porous media containing both oil and water. *Soc Pet Eng J*. 1969;9:3–12.
3. Foster WR. A low tension waterflooding process employing a petroleum sulfonate, inorganic salts, and a biopolymer. *SPE AIME Mid-Continent Sect Improved Oil Recovery (United States)*, 1972:25.
4. Delshad M. *UTCHEM VERSION6.1 Technical Documentation*. Austin, TX: Center for Petroleum and Geosystems Engineering, the University of Texas at Austin; 1997.
5. Qi LQ, Liu ZZ, Chen L, et al. Discovery of two different driving conditions in combination flooding process by numerical simulation. *Petrol Geol Oilfield Dev Daqing*. 2009;28:84–90.
6. Qi LQ, Shi Y, Wang HS, et al. Two different driving conditions in combination flooding process. *Petrol Geol Oilfield Dev Daqing*. 2009;28:91–97.
7. Harbert LW. *Low interfacial tension relative permeability*. SPE 12171, 1983.
8. Bardon C, Longeron D. Influence of very low interfacial tensions on relative permeability. SPE 7609, *SPE J*. 1980;20:391–401.
9. Ronde H. *Relative permeability at low interfacial tension*. SPE 24877, 1992.

10. Liu FH, Huang YZ. Influences of interfacial tension on relative permeability curve. *Petrol Geol Oilfield Dev Daqing*. 2002;24:50–52.
11. Lu GQ, Wang YD, Chen YM, et al. Experimental study on influences of low interfacial tension system on relative permeability. *Oilfield Chem.* 2003;20:54–57.
12. Ye ZB, Peng KZ, Wu XL, et al. Study on ASP flooding relative permeability curve for Karamay oilfield. *J Petroleum.* 2000;21:49–54.

Chapter 2

Digital research on field experiment of combination flooding

Lianqing Qi[a,b], Hongshen Wang[b], Kaoping Song[c], Quangang Liu[b], Jinlin Wang[b], Tao Yu[d], Xiaochao Wang[b], and Si Shen[b]
[a]*CNPC Daqing Oilfield Exploration and Development Research Institute, Daqing, China,* [b]*CNOOC Energy Technology & Services-Drilling & Production Technology Services Co., Tianjin, China,* [c]*Northeast Petroleum University, Daqing, China,* [d]*CNPC Liaohe Oilfield Exploration and Development Institute, Panjin, China*

Chapter outline

Development and Application of Classical Capillary Number Curve Theory
https://doi.org/10.1016/B978-0-12-821225-7.00002-3

The United States begun researching combination flooding field experiments in the 1980s and launched flooding experiments at the West Kiehl Oilfield in Wyoming,[1] and alkali-surfactant-polymer (ASP) flooding field experiments at the Cambridge Oilfield in Ohio[2] and in the Warden Unit of the Sho-Vel-Tum Oilfield in Oklahoma.[3] These experiments were closely followed by tests at Daqing Oilfield in China. On the basis of sound indoor researches, Daqing Oilfield started pilot tests in the 1990s, moved on to expansive tests after initial success, and then expanded to industrial application field experiments. Wang et al.[4] give a detailed description of these field experiments. In recent years, Shengli Oilfield in China launched surfactant and polymer (SP) flooding field experiments. The flooding experiments are valuable in the study of combination flooding technology.

Qi et al.[5] introduced the research and development of capillary number test curve *QL* and Improved Mechanism of Combination Flooding (IMCFS) software. The test curve *QL* perfected the understanding of the flooding mechanism of combination flooding, and the IMCFS software offered effective means for researching and summarizing the field experiments of combination flooding. Numerical simulation methods were used to research domestic and overseas field experiments, including those at Daqing Oilfield, check and assess the new achievements of combination flooding theoretical studies, summarize the experiences and lessons of field flooding tests, and put forward more accurate and reasonable flooding schemes so as to promote research into and applications of combination flooding technologies.

2.1 Numerical simulation research of Xingerxi Block Area (simplified Xinrerxi Block Area) combination flooding experiment of No. 4 oil production plant in Daqing Oilfield

2.1.1 Research on numerical simulation calculation of Xingerxi Block Area combination flooding experiment

Wang et al.[6] provided the technical data for the combination flooding field experiments at Xingerxi Block Area, Daqing Oilfield. The well site schematic

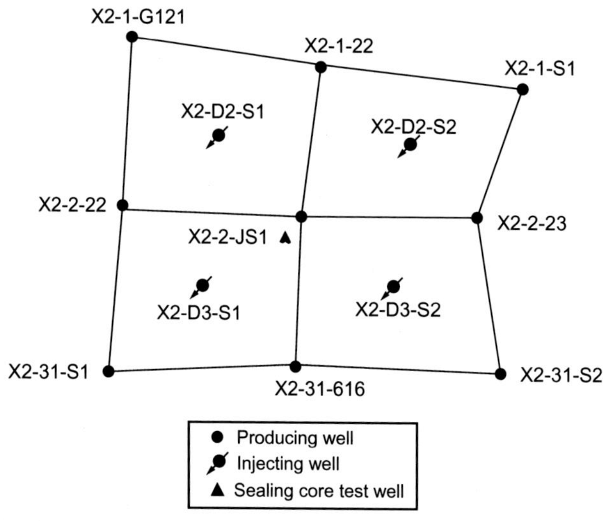

FIG. 2.1 Sketch of well location of ASP combination flooding test AREA of Xingerxi Block Area.

diagram is shown in Fig. 2.1: the test area was 0.3 km^2, sandstone thickness was 7 m, the effective thickness was 5.8 m, the effective permeability was 0.675 μm^2, and the reservoir permeability coefficient of variation was 0.65. There were four injection wells and nine production wells in the whole area, with well spacing of 200 m. Table 2.1 displays the implementation scheme and detailed data. Wang et al.[6] confirmed that the underground system interfacial tension of the flooding experiment was 1.25×10^{-3} mN/m or so, operating viscosity was about 30 mPa s, and annual injection speed of flooding experiment was 0.24 PV.

TABLE 2.1 Implementation table of Xingerxi Block Area combination flooding test scheme.

Block	Preslug	Main slug	Auxiliary slug	Subsequent polymer slug		
				1	*2*	*3*
Volume (PV)	0.0376	0.351	0.10	0.05	0.10	0.05
Alkali (%)		1.2	1.2			
Surfactant concentration (%)		0.3	0.1			
Polymer concentration (mg/L)	1500	2300	1800	1000	700	1000

The simplified geologic structure model in Qi[7] was used to describe the oil reservoirs of different heterogeneous coefficients of variation: homogeneous reservoir plane, vertical heterogeneous three-layer structure. The heterogeneous coefficients of variation were different and the corresponding interval had corresponding permeability. The different permutations and combinations determined the different deposition patterns of the oil reservoir.

The flooding experiment at Xingerxi Block Area was a preliminary test, in which a five-point method was used for well pattern distribution of "one injection well and four production wells." The structure calculation model shown in Fig. 2.2 was used in simulation calculation studies, which belonged to one-fourth of the well group in a well pattern of five-point methods, including one injection well and one production well. In order to fully reflect the change conditions of chemical substances in the oil reservoir and the flowing state of the fluid, 9×9 well grids were taken on the plane and oil/water wells were separated by seven grids.

The IMCFS software was chosen for the fitting calculation research of the flooding experiment at Xingerxi Block Area. The related parameters required in the calculation, including "capillary number test curve *QL*" and "phase permeability curve *QL*," were all determined and measured via the tests under the oil/water conditions of No. 4 Oil Production Plant of Daqing Oilfield.

The flooding experiment at Xingerxi Block Area was carried out under the ultra-high water cut condition. Firstly, we fitted the flooding process and adjusted slightly the permeability of oil reservoir stratification and the related data of phase permeability curve—saturation of stratified flooding residual oil—and fitted the flooding recovery target. The recovery percentage was 47.2% when the moisture content of the produced fluid from the oil well was 98%, which is consistent with the production data on the site. We determined the basic physical parameters of the oil reservoir, and the

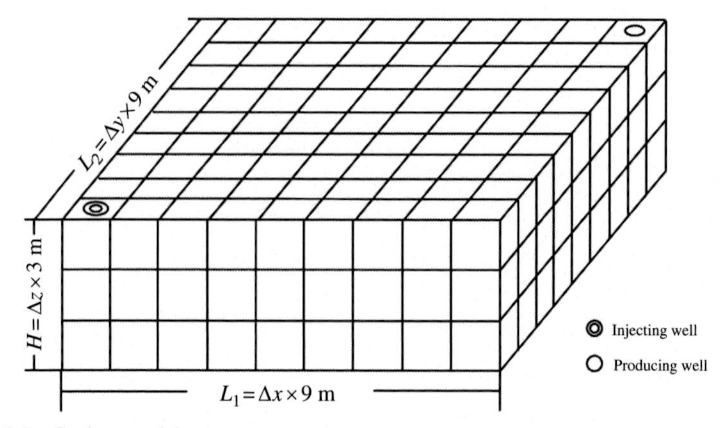

FIG. 2.2 Geology model structure.

permeability of three intervals were corrected as 100, 215, and 525 $\times$ 10^{-3} μm^2. We continued to inject the water until the water cut of the oil well reached 99.82% and the production reached 52.8%, and then we injected the polymer preslug. The combination flooding process began. The saturation parameter S_{or}^{H} of residual oil and polymer solution viscosity-concentration curve shear rate of the three intervals were amended to fit the curve between change of oil-well water cut and change of enhanced oil recovery. Fig. 2.3 displays the on-site flooding experiment and the change curve of both water cut and recovery of oil wells resulted from the fitting calculation, whose heights were consistent. The fitting of flooding experiment achieved a satisfactory result.

Table 2.2 shows the distribution data of low-permeability reservoir residual oil saturation and surfactant concentration after the end of the flooding experiment. It can be seen that there was a relatively low oil-containing saturation only on nearly half a level in the side of water well in the upper left corner. There was high oil-containing saturation on the grid in a large area in the front, where the "oil wall" gathered here by the crude oils flooded from the rear area. According to the corresponding surfactant concentration distribution, it can be seen that the position of the "oil wall" corresponded to high surfactant concentration, where the main body of the composite system slug finally resided.

Note in particular that there was a higher oil-containing saturation in the "second" grid behind the oil well on the mainstream line compared to the adjacent grid, which is the "peak point" of the oil wall. Li et al.[8] introduced that after the end of the flooding experiment at Xingerxi Block Area, high-saturation crude oil was found in the core of the closed coring well "Xing 2-2—Inspection Well No. 1" 50 m away from the oil well on the mainstream line, and the inspection well was exactly at the "peak point" of oil well under simulated calculation.

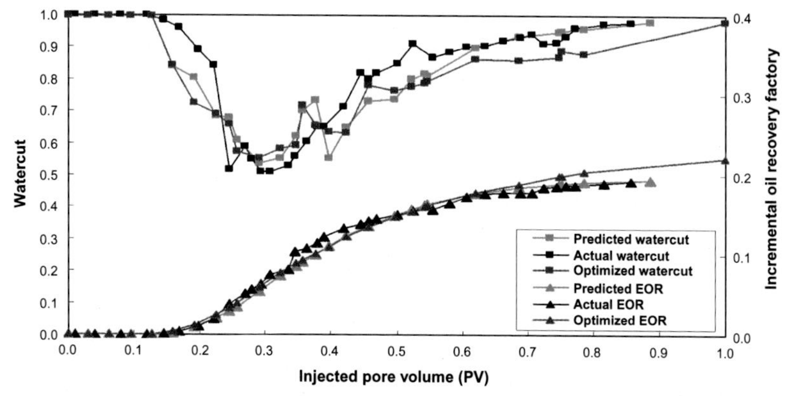

FIG. 2.3 Water cut and recovery curves of Xingerxi Block Area combination flooding test, scheme matching and optimization scheme.

TABLE 2.2 Remaining oil distribution table of low permeability layer at the end of Xingerxi Block Area combination flooding test (%).

	Xingerxi Block Area	1	2	3	4	5	6	7	8	9
Residual Oil %	1	15.7	15.8	16.0	16.2	16.8	19.2	34.2	39.5	50.6
	2	15.8	15.9	16.0	16.3	16.8	19.4	32.8	39.8	50.3
	3	16.0	16.0	16.2	16.5	17.0	20.5	32.6	40.4	49.7
	4	16.2	16.3	16.5	16.8	17.5	23.2	33.6	41.4	49.5
	5	16.8	16.8	17.0	17.5	19.8	25.9	35.9	42.4	49.2
	6	19.1	19.3	20.4	23.1	26.0	32.0	40.2	43.7	48.9
	7	34.0	32.6	32.4	33.4	35.9	40.9	44.0	43.6	48.6
	8	39.5	39.8	40.3	41.2	42.4	43.7	43.6	44.0	47.3
	9	50.6	50.3	49.6	49.4	49.2	48.9	48.7	47.4	47.8
Surfactant Concentration %	1	0.001	0.002	0.01	0.03	0.063	0.115	0.131	0.113	0.037
	2	0.002	0.005	0.017	0.032	0.073	0.116	0.132	0.117	0.053
	3	0.010	0.018	0.031	0.043	0.089	0.122	0.133	0.121	0.061
	4	0.031	0.032	0.045	0.079	0.099	0.116	0.125	0.123	0.066
	5	0.06	0.073	0.087	0.100	0.098	0.090	0.113	0.125	0.071
	6	0.110	0.117	0.119	0.115	0.088	0.085	0.103	0.124	0.079
	7	0.131	0.132	0.128	0.12	0.107	0.098	0.112	0.122	0.08
	8	0.114	0.118	0.12	0.123	0.124	0.120	0.119	0.116	0.082
	9	0.038	0.056	0.064	0.071	0.074	0.079	0.082	0.081	0.065

The simulated calculation understood the causes for a high-saturation crude oil existed in the core of inspection well.

The calculation results showed that the experiment terminated when water cut in produced fluid of the oil well rose to 98% after a sharp decline, and the recovery percentage was 72.45%, which increased by 19.65% compared with the moment of injection transfer and was consistent with the actual field condition, and increased by 25.25% compared with the moment when flooding oil well water cut was 98%.

It is commonly considered that "the recovery percentage at the time when water cut in produced fluid of oil well rose to 98% after a sharp decline in the injection transfer system is determined as the recovery percentage of combination flooding, and its difference from the recovery percentage at the moment when the water cut of flooding oil well is 98% is served as the enhanced recovery percentage and can also be called 'enhanced recovery percentage'." According to this definition, the enhanced oil recovery in the experiment should be 25.25%. Wang et al.[4] took "the difference between the final recovery percentage of combination flooding and the initial experiment flooding recovery percentage as the enhanced oil recovery of flooding experiment" and determined that the percentage of enhanced oil recovery in this experiment was 19.6%. Two targets of enhanced oil recovery were acquired from the definition of double standards, which makes it difficult to correctly evaluate the experiment. Therefore it is necessary to unify them under a common definition. According to the common definition, the percentage of enhanced oil recovery in the experiment at Xingerxi Block Area should be 25%. Hence the experiment of Xingerxi Block Area is an early flooding test with an ideal effect of flooding.

2.1.2 Digital flooding experiment research platform and digital flooding experiment

The acquisition of satisfactory fitting results from fitting calculation thereby determined the reservoir geology data and corresponding information data of combination flooding and established the digital geology model platform of reservoir combination flooding data. The flooding scheme can be calculated via the digital geology model platform and the final scheme of fitting test calculation can also be regarded as the first flooding scheme calculated on the geology model platform. There is a relatively high accuracy based on the scheme calculation. The calculation verified that the flooding schemes calculated on this platform all had a relatively high accuracy. The high-accuracy results based on the flooding scheme calculation could regard flooding scheme calculation as "digital flooding experiment." Because the IMCFS software adopted "capillary number test curve *QL*" and the corresponding phase permeability curve and showed the detailed description of the relationship between capillary number and residual oil saturation under the condition of high capillary number, based on the data information acquired from the digital flooding experiments, we could more clearly understand the flooding experiment, deepen the researches of digital flooding experiments, obtain efficient and feasible flooding schemes, and promote the research and application of combination flooding technology.

2.1.3 Important technical characteristics for success of Xingerxi Block Area flooding experiment—High-viscous and ultra-low interfacial tension combination system

The fitting calculation process of the Xingerxi Block Area flooding experiment confirmed that the maximum capillary number in local scope of the reservoir's high-permeability layer during the experiment reached 0.1687, higher than extreme capillary number N_{ct1} 0.072. This indicated that the flooding process was in a "class-II" flooding state. By enhancing the interfacial tension of the experiment scheme system to calculate the comparison scheme and controlling the maximum capillary number of the oil reservoir to be slightly lower than extreme capillary number N_{ct1} 0.072 during the flooding process, the flooding process was then in the "class-I" flooding state. Table 2.3 lists the relevant parameters and the flooding results of the experiment scheme and comparison scheme. From them, we learn that the three intervals of the experiment scheme all had relatively low residual oil saturation, and had, as a whole, a relatively high percentage of recovery efficiency.

TABLE 2.3 Characteristic parameters and displacing effects contrasting table of different schemes.

Scheme	Implementation time (d)	Parameter of combinational system			Stratified residual oil for scheme termination (%)			Oil recovery (%)	Increased recovery factor (%)	Oil increase (t)	Type of flooding state
		Interfacial tension 10^{-3} (mN/m)	*Maximum viscosity (mPa s)*	*Maximum capillary number*	*Upper layer*	*Middle layer*	*Lower layer*				
Experiment	1567	1.25	31.04	0.1687	31.92	17.96	14.32	72.45	25.25	16307	II
Comparison	1507	3.0	30.06	0.0701	33.18	18.98	14.50	71.39	24.19	15622	I
Optimization	1450	1.25	22.93	0.1121	28.87	17.51	14.45	73.90	26.70	17243	II

Table 2.4 shows the reservoir grid capillary numbers of high-permeability layers for the experiment scheme and comparison scheme and the distribution data of residual oil saturation at the termination moment of schemes. Moment A refers to the time when the injection volume of combinational system reaches up to 0.345 PV. It could be seen that a small area marked with "red ink" appeared near high-permeability layer water well in the flooding experiment, and the value of these grid capillary numbers was higher than

TABLE 2.4 Reservoir grid capillary number and residual oil saturation distribution table for different schemes of Xingerxi Block Area at different time.

Xingerxi Block Area		1	2	3	4	5	6	7	8	9
Capillary Number of High-Permeability Reservoir at Moment A of Experiment Scheme	1	**0.1687**	**0.0881**	0.0565	0.0403	0.0294	0.0194	0.0080	0	0
	2	**0.0881**	0.0688	0.0513	0.0388	0.0289	0.0192	0.0071	0	0
	3	0.0565	0.0513	0.0429	0.0347	0.0267	0.0176	0.0009	0	0
	4	0.0403	0.0388	0.0347	0.0295	0.0226	0.0116	0	0	0
	5	0.0294	0.0289	0.0267	0.0226	0.0177	0.0003	0	0	0
	6	0.0194	0.0192	0.0176	0.0116	0.0003	0	0	0	0
	7	0.0081	0.0072	0.0009	0	0	0	0	0	0
	8	0	0	0	0	0	0	0	0	0
	9	0	0	0	0	0	0	0	0	0
Capillary Number of High-Permeability Reservoir at Moment B of Experiment Scheme	1	0.0031	0.0304	0.0385	0.0305	0.0248	0.021	0.0155	0.0088	0.0012
	2	0.0304	0.042	0.0371	0.0304	0.026	0.0224	0.0173	0.0116	0.0017
	3	0.0391	0.0374	0.0331	0.0294	0.027	0.0238	0.0194	0.0147	0.0017
	4	0.0313	0.031	0.0297	0.029	0.0276	0.0248	0.0213	0.0172	0.001
	5	0.0259	0.027	0.0278	0.0281	0.0273	0.0255	0.0232	0.0198	0.0003
	6	0.0223	0.0237	0.0252	0.026	0.0263	0.0262	0.0247	0.0237	0.0001
	7	0.0169	0.0188	0.021	0.0232	0.0252	0.026	0.0262	0.0164	0.000
	8	0.0098	0.0129	0.0165	0.0201	0.0233	0.0245	0.0233	0.0045	0.000
	9	0.0014	0.0016	0.0019	0.0014	0.0007	0.0004	0.0005	0.0005	0.0005
Residual Oil Saturation of High Permeability Reservoir at Termination Moment of Experiment Scheme %	1	4.7	13.9	14.1	14.2	14.1	14.0	14.0	14.3	16.8
	2	13.9	14.0	14.2	14.2	14.0	13.9	14.0	14.2	15.9
	3	14.1	14.2	14.3	14.1	13.9	13.9	14.0	14.2	16.1
	4	14.2	14.2	14.1	13.9	13.9	13.9	14.0	14.2	16.2
	5	14.1	14.0	13.9	13.9	13.9	13.9	14.0	14.2	16.1
	6	14.0	13.9	13.9	13.9	13.9	13.9	14.0	14.2	15.9
	7	14.0	14.0	14.0	14.0	14.0	14.0	14.0	14.1	15.6
	8	14.2	14.2	14.2	14.2	14.2	14.1	14.1	14.1	15.2
	9	16.7	15.9	16.1	16.1	16.0	15.7	15.4	15.0	15.4
Capillary Number of High-Permeability Reservoir at Moment B of Comparison Scheme	1	0.0025	0.016	0.0163	0.0129	0.0104	0.0088	0.0064	0.0037	0.0008
	2	0.016	0.0179	0.0157	0.0128	0.0109	0.0093	0.0072	0.0048	0.0013
	3	0.0166	0.0159	0.014	0.0123	0.0113	0.010	0.008	0.006	0.0015
	4	0.0133	0.0131	0.0125	0.0122	0.0116	0.0104	0.0088	0.0071	0.0008
	5	0.011	0.0114	0.0117	0.0118	0.0115	0.0107	0.0095	0.008	0.0003
	6	0.0094	0.01	0.0106	0.011	0.0111	0.0109	0.0103	0.0097	0.0001
	7	0.0071	0.0079	0.0089	0.0099	0.0105	0.0109	0.011	0.0088	0.0000
	8	0.0041	0.0054	0.007	0.0085	0.0098	0.0103	0.0104	0.0039	0.0000
	9	0.001	0.0016	0.002	0.0013	0.0006	0.0004	0.0004	0.0004	0.0004
Residual Oil Saturation of High Permeability Reservoir at Termination Moment of Comparison Scheme %	1	13.5	13.6	13.6	13.7	13.8	13.8	14.0	14.5	24.4
	2	13.6	13.6	13.6	13.7	13.8	13.8	14.0	14.3	16.6
	3	13.6	13.6	13.7	13.7	13.8	13.9	14.0	14.2	15.6
	4	13.7	13.7	13.7	13.8	13.8	13.9	14.0	14.2	15.6
	5	13.8	13.8	13.8	13.8	13.9	13.9	14.0	14.2	15.9
	6	13.8	13.8	13.9	13.9	13.9	13.9	14.0	14.2	15.9
	7	14.0	14.0	14.0	14.0	14.0	14.0	14.0	14.2	15.7
	8	14.5	14.3	14.2	14.2	14.2	14.2	14.1	14.2	15.3
	9	24.1	16.4	15.6	15.5	15.8	15.8	15.5	15.1	15.7

the value 0.072 of extreme capillary number N_{ct1} and the grids were in the "class-II" flooding state. They correspond to relative high residual oil saturation, relatively rapid water-phase flow rate, and a poor displacement effect. However, a large area of grids was marked by "green ink" in the front. The numerical range of capillary numbers was between maximum capillary number N_{ct2} and extreme capillary number $N_{ct1,}$ and they correspond to "residual oil saturation S_{or}^{H}," which are in range of capillary numbers under "class-I" flooding state, with the best displacement effect. The aforementioned positions were influenced by the combinational system, in front of which the area marked by "blue-green ink" is the frontal area without influence by combinational system slug, and its capillary number is extremely low. Moment B refers to the time when the injection volume of combinational system reaches up to 0.460 PV. The "class-II" flooding state grids of the reservoir's high-permeability layer disappeared, which indicated that the "class-II" flooding state did not last for a long time. The vast majority of grids were in the "class-I" flooding state with the best displacement effect. By comparing with the data of the reservoir's high-permeability layer grid capillary number in the comparison scheme, this scheme did not experience the "class-II" flooding state. Through comparative analysis of capillary numbers at moment B, it could be seen that the capillary number ranges of two schemes were basically the same during the best flooding effect while they were in the "class-I" flooding state. Based on the numerical comparison, there generally was a relatively big number for the grids corresponding to the experiment scheme, and it was quite significant especially for the grids at two sides of mainstream line, which showed that the experiment scheme displaced and expanded the repercussion effect under the condition of high capillary number. Through analyzing the distribution data of residual oil saturation of high-permeability interval grids at the termination of scheme, the comparison schemes all had a low residual oil saturation and good displacement effect in the range of large area. The experiment scheme of Xingerxi Block Area was for a short time displaced under the "class-II" flooding state with high capillary number and still flooded by "class-I" flooding state into overall flooding effect of main control. The expanded repercussion effect displaced by its "class-II" flooding state relatively improved the displacement effect of positions at two sides of the mainstream line, which has the relatively good flooding effect.

The researchers concluded that the application of an ultra-low interfacial tension, high-viscous system in the flooding experiment guaranteed the acquisition of relatively higher enhanced effect of recovery efficiency.

2.1.4 Further optimization of flooding experiment scheme of Xingerxi Block Area

Table 2.2 lists the oil wall position detained in the low-permeability reservoir and the main-body position of the combinational system solution slug at the moment of experiment termination. To achieve the goal of residual oil in low-permeability reservoir, the polymer slug viscosity behind the main-body slug could not be lower than the viscosity of the main-body slug and the slug must have enough length. This is the only way to inhibit the fingering of high-permeability layer and maintain and continue the good flooding trend of low-permeability layer and ensure more oil are produced from the low-permeability positions or layers so as to obtain a better displacement effect.

Based on this analysis and with reference to America's oilfield experiment schemes,[1, 2] the "two-stage structure" optimal model flooding scheme was designed: a combinational system slug with a volume of 0.3 PV and surfactant concentration of 0.3%, subsequent polymer slug with a volume of 0.558 PV and polymer concentration of 1830 mg/L (the same for two-stage slug). Calculated based on that surfactant price is 1.5 times the polymer price, the expense of chemicals for this scheme is basically the same as that of field experiment scheme calculated. Table 2.3 lists the main technical and economic indicators of the optimal model scheme of Xingerxi Block Area. The comparison of experiment scheme data indicated that the percentage of enhanced oil recovery by the optimal scheme was higher than 1.45% and oil production increased by 936 tons as a whole. The oil increase was 4 tons for ton equivalent polymer. This scheme can be considered as a relatively optimal flooding scheme. Fig. 2.3 shows the change curves of both oil well water cut decline and enhanced recovery percentage in the optimal scheme, which indicates the enhanced recovery effect of the optimal scheme. After the displacement of optimal "model" flooding scheme, there was still more than 28% residual oil in the low-permeability reservoir and there still was a potential of enhanced recovery.

2.1.5 Some noteworthy technical data

2.1.5.1 Safe injection pressure limit

The high-viscosity system injected on the site of combinational flooding experiment has certain safe injection pressure limit P. The polymer concentration of combinational system main slug injected safely on the experiment site was up to 2300 mg/L, and after the end of fitting flooding experiment process, the maximum average pressure P1 of the oil reservoir after completing the injection of high-viscosity combinational system slug and polymer slug was obtained, which corresponds to safe injection pressure limit P on the site. P1 was then determined as the simulated and calculated injection pressure limit.

2.1.5.2 Underground viscosity retention rate of flooding system and economic indicators of enhanced recovery for flooding experiment

The "viscosity retention rate" of viscosity-concentration relationship curve selected in the fitting calculation of Xingerxi Block Area was 30%, and a satisfactory fitting result was acquired. It was confirmed that the underground viscosity retention rate of polymer solution for field experiment was 30%. The determination of underground viscosity retention rate of polymer solution guaranteed the consistency of polymer consumptions for both scheme calculation and field experiment. We took the surfactant price as 1.5 times the polymer price and converted the surfactant consumption into the equivalent polymer consumption, obtained via the calculation the equivalent polymer consumption of overall consumption of chemicals and the "ton equivalent polymer oil increase" of economic indicators of evaluating the combination flooding enhanced effect of recovery. Based on the experiment of Xingerxi Block Area with a good flooding effect, we obtained 60.13 tons of "oil increase for ton equivalent polymer" and determined "60 tons of oil increase for ton equivalent polymer" as the reference standard of combination flooding experiment with good economic effect.

2.1.5.3 Requirements on stabilizing time of combination flooding surfactant

According to the distribution data of surfactant concentration of low-permeability reservoir listed in Table 2.2, the front of main-body part of the combinational system slug just arrived at the oil well at the moment of scheme termination, and the slug main-body part still detained in middle and front position of low-permeability interval (reservoir). The fitting calculation was carried out under the condition of stable chemicals. Attention was paid to analyzing the condition of fitting calculation of oil-well water cut change curves. The two curves quite ideally coincided in the later period. The good coincidence of change curve and calculation curve of oil-well water cut for flooding experiments indicated that no failure of surfactant occurred in the later period of the experiment. Otherwise, the water cut rising of oil well and the declination of test effect would accelerate the upward situation of the curve due to surfactant failure, and there was no abnormity for the change curve of enhanced oil recovery. After analyzing the change condition of surfactant concentration of oil reservoir during the flooding process, it could be seen that the surfactant solution was gradual. If the stabilizing time of surfactant solution was beyond the actual effect (failure) in a certain position at a certain moment, the subsequently arrived solution was bound to result in timeout failure. For this reason, it was necessary to make sure there was no failure of surfactant solution during the whole experiment process. The implementation time of the flooding experiment of Xingerxi Block Area was approximately 1570 days. The experiments proved that the surfactant system could maintain the stability under the ground in such a time. According to the experiment results, the requirements on stabilizing time limit of experiment surfactants was defined, namely, it could not exceed 1570 days.

2.2 Optimization design of flooding experiment scheme of combination flooding

On the digital geology model platform of Xingerxi Block Area flooding experiment, the optimization design of flooding experiment scheme was subject to an in-depth study. The scheme result is listed in Table 2.5.

2.2.1 Optimization design of well pattern and spacing

The practice of Daqing Oilfield proved that the five-point well pattern was the well pattern applicable to the chemical flooding, and its key lied in that each production well was benefited from the flooding effect of four injection wells of the combinational system. Producer-injector spacing is an important issue in the scheme design of combination flooding. Under the condition that the producer-injector spacing was 250 m, we calculated flooding Scheme 1.1. The combinational system slug volume was 0.3 PV, surfactant concentration was 0.3%, subsequent polymer slug volume was 0.2 PV, and maximum average pressure of oil reservoir during the oil flooding process was lower than pressure limit P1. The chemical flooding of Daqing Oilfield adopted the uniform liquid-injection speed—the injection strength was equivalent to that under the condition of producer-injector spacing being 250 m and injection speed being 0.15 PV/y, hereafter called "the first injection speed." The first injection speed was adopted in this scheme. From the table, we learn that the biggest problem of this scheme is its implementation time of 2260 days, which is nearly two years longer than the time required by the scheme implementation. We adjusted the slug volume and calculated flooding Scheme 1.2. From the table, it can be seen that the implementation time of the scheme was shortened to 1561 days, which met the system stability requirement and the percentage of enhanced recovery decreased to 13.5%. Further, the producer-injector spacing was shortened to 125 m and calculated the flooding Scheme 1.3. In addition, the implementation time of the scheme was attained and the recovery percentage rose to 27.87%.

The information from the site at Daqing Oilfield indicates that 125 m of producer-injector spacing obtained a good effect. Based on the theoretical studies and field effect, it is recommended to take the appropriate small spacing for the research and application of combination flooding technology.

2.2.2 Design of slug structure for flooding scheme

Wang et al.[4] showed that the polymer preslug was provided in front of slug of testing combinational system after the experiment at Xingerxi Block Area for purpose of "the polymer preslug has 'profile control' effect to reinforce the entry of combinational system in low-permeability reservoir and improve

TABLE 2.5 Calculation results of optimized flooding schemes.

Scheme	Slug volume (PV)		Imple-menta-tion time (d)	Maximum under-ground viscosity (mPa s)	Reservoir maximum average pressure ratio	Recovery percent (%)	Enhanced recovery percent-age (%)	Stratified residual oil (%)			Surfactant consump-tion (t)	Equivalent of total chemicals consump-tion P (t)	Oil increase for ton equi-valent polymer (t/t)
	Combi-national system	*Poly-mer*						*Upper layer*	*Middle layer*	*Lower layer*			
1.1	0.30	0.20	2260	32.45	1.010	69.19	22.03	35.58	21.09	15.13	86.32	296.8	73.84
1.2	0.15	0.18	1561	39.34	0.871	60.66	13.50	40.28	30.91	20.47	32.81	134.9	99.56
1.3	0.30	0.60	785	25.67	0.992	75.11	27.87	27.10	16.75	14.15	31.93	111.1	62.34
2.1	0.30	0.60	783	25.62	0.985	74.84	27.60	27.68	16.81	14.14	31.93	112.0	61.24
3.1	**0.30**	**0.60**	**1170**	**33.35**	**0.997**	**76.74**	**29.50**	**24.39**	**15.92**	**13.88**	**32.00**	**123.2**	**59.50**
4.1	0.20	0.70	1100	31.16	0.997	74.87	27.63	27.69	16.78	14.09	21.29	103.3	66.47
4.2	0.45	0.45	1133	34.70	0.995	77.56	30.32	22.61	15.86	13.82	47.97	149.5	50.40
5.1	0.45	0.35	1123	33.26	0.995	75.61	28.37	26.77	16.17	13.88	31.98	123.1	57.27
5.2	0.20	0.70	1153	32.77	0.994	77.09	29.85	23.56	15.92	13.92	31.94	122.4	60.60
6.1	0.30	0.60	1115	32.23	0.998	74.32	27.08	28.25	17.42	14.17	32.00	121.1	55.57
6.2	0.30	0.60	1048	31.98	0.999	70.61	23.37	32.10	21.56	14.83	32.00	120.8	48.07
7.1	0.30	0.50	1053	34.80	0.990	76.00	28.76	25.99	16.06	13.87	32.00	117.2	60.98
7.2	0.30	0.70	1220	32.02	0.995	77.08	29.84	23.57	15.89	13.93	32.00	129.0	57.48

the flooding effect." Such slug structure design, however, did not appear in the United States. On the basis of Scheme 1.3, the preslug with a volume of 0.05 PV was added in front of the combinational system slug to calculate the flooding Scheme 2.1. The calculation results indicated that the flooding effect was less than that of Scheme 1.3. The simulated calculation studies showed that the enhanced recovery effect of polymer preslug was not significant. If no polymer preslug was added and the polymer saved was used for the subsequent "protective" slug, a better flooding effect could be obtained. Therefore, it is not recommended to set up the polymer preslug for the flooding scheme.

2.2.3 Design of injection speed for flooding scheme

The in-depth studies showed that transfusion speed and system viscosity had different contributions during the flooding process. The contribution of the transfusion speed only had the displacement role. The enhancement of the system viscosity could not only inhibit the displacement fluid to press onward to expand sweep volume and improve the displacement effect of two sides of plane mainstream line of oil reservoir and low-permeability layer, but also improve the oil and water mobility ratio and benefit the oil flow and production, which made the outstanding contributions to the enhancement of flooding effect. We calculated flooding Scheme 1.3 by taking the "the second injection speed" – 2/3 × "the first injection speed." The calculation results were detailed in Scheme 3.1. It can be clearly seen from the table that the maximum underground operating viscosity was 33.35 mPa s, which enhanced more than 7 mPa s compared to that in Scheme 1.3. The recovery efficiency increased by 1.63% and the implementation time was 1170 days, which meets the stability requirements and has been prolonged compared to that in Scheme 1.3. As far as the scheme of small well spacing is concerned, there are some advantages for no difficulty, which enables the equipment to be fully used. It is recommended to use "the second injection speed" under the condition of optimizing the flooding scheme with a relatively small well spacing.

2.2.4 Optimization of slug volume of combinational system

On the basis of Scheme 3.1, the two schemes for calculation of slug volume of the combinational system were adjusted. The slug volume of Scheme 4.1 was 0.20 PV and the slug volume of Scheme 4.2 was 0.45 PV. From the table, we learn that the slug of combinational system in Scheme 4.1 reduced 0.10 PV and the recovery percentage decreased by 1.87% compared to that in Scheme 3.1. The slug of the combinational system in Scheme 4.2 increased by 0.15 PV and the recovery percentage increased by 0.82% compared to that in Scheme 3.1. It was clear that the relative enhanced recovery effect of the latter scheme was not good. It is noted that the excessive consumption of expensive surfactants in Scheme 4.2 resulted in the increased cost, with only 50.40 tons of oil increase for ton equivalent polymer. The studies showed that the optimized volume of the combinational system slug was 0.30 PV or so.

2.2.5 Optimization of surfactant concentration of combinational system slug

The two schemes for calculation of combinational system slug volume and surfactant concentration in Scheme 3.1 were adjusted simultaneously. The slug volume in Scheme 5.1 was 0.45 PV and surfactant concentration was 0.2%. The slug volume in Scheme 5.2 was 0.20 PV and the surfactant concentration was 0.45%. The surfactant volume consumptions of the combinational system were all 900 mg/L·PV in the three schemes. The enhanced recovery percentage in Scheme 5.1 decreased by 1.13% compared to that in Scheme 3.1, and the enhanced recovery percentage in Scheme 5.2 increased by 0.35% compared to that in Scheme 3.1. We can learn from this that the enhanced recovery effect of the scheme with high surfactant concentration is relatively good, which resulted from the high diffusion effect of the concentration. It is recommended to take the scheme of slug surfactant concentration of 0.30% that has been successfully assessed by a large number of tests, and take the scheme of surfactant concentration of 0.45% as the priority assessment scheme.

2.2.6 Optimization of interfacial tension of combinational system slug

The combinational system interfacial tension of Scheme 3.1 was 0.00125 mN/m. We calculated Schemes 6.1 and 6.2 by changing the system's interfacial tension as 0.005mN/m and 0.0095mN/m, respectively. The enhanced recovery percentage in Scheme 6.1 decreased by 2.42% compared to that in Scheme 3.1, and the enhanced recovery percentage in Scheme 6.2 decreased by 6.13% compared to that in Scheme 3.1. The system interfacial tensions of the three schemes all met the technical standards of domestic petroleum industry—the system interfacial tension of the combinational system shall be less than 10^{-2} mN/m. The difference of the enhanced recovery percentages was so great that more attention should be paid to the fact that the crude oil not produced might be discarded under the ground for good. It was necessary to set high standard requirements and try as best to use the relatively lower interfacial tension system for oil flooding, which will not only achieve a higher target of enhanced oil recovery but also ensure the crude oil with a lower value to be left under the ground.

2.2.7 Volume optimization of subsequent polymer slug

The two schemes for calculation of subsequent polymer slug volume in Scheme 3.1 were changed. The subsequent polymer slug volume in Scheme 7.1 was 0.4

PV, and the enhanced recovery percentage in Scheme 7.1 decreased by 0.74% compared to that in Scheme 3.1, with 60.98 tons of oil increase for ton equivalent polymer, higher than that in Scheme 3.1. The subsequent polymer slug volume in Scheme 7.2 was 0.7 PV and the enhanced recovery percentage in Scheme 7.2 was 29.84%, higher than that in Scheme 3.1. Implementation time was 1753 days, longer than half a year, with 57.48 tons of oil increase for ton equivalent polymer.

The optimization of flooding scheme shall be "the optimal scheme of enhanced recovery target shall be taken under the condition that the production technology is safe and feasible and the economic indicator can be accepted." According to the guiding ideology, Scheme 3.1 is recommended as the optimization scheme.

2.3 Research of flooding experiment for industrial well pattern condition

2.3.1 Digitization of industrial experiment of No. 4 oil production plant in Daqing Oilfield

Daqing Oilfield launched the industrial field experiments after successfully implementing the pilot test, the enlargement test of combination flooding, and other preliminary tests. Wang et al.[4] introduced the industrial field experiments in Daqing Oilfield, including experiments for Middle Xingbei Block 2, Southern Block 5, North Block 1 East Fault, North Block 3 West, and others. Some experiments were under way when the literature was published, so the enhanced oil recovery provided by the literature was the predicted value (Table 2.6).

Firstly, a noteworthy problem was found via the experiment at Middle Xingbei Block 2. This experiment and the experiment at Xingerxi Block Area were all conducted in Block 2 of the No.4 Oil Production Plant and the target reservoirs for both experiments were adjacent. The experiment of Middle Xingbei Block 2 was the expansive test based on the experiment of Xingerxi Block Area. The slug polymer concentration of the combinational system in the experiment at Xingerxi Block Area was 2300 mg/L, and the slug polymer concentration in the experiment of Middle Xingbei Block 2 was only 1000 mg/L. It was understood from the field that there was a high polymer concentration designed in the original scheme, and it was forced to be changed into a low concentration due to the injection difficulty during the implementation process. It was not difficult to find that there was a significant difference of polymer slug injection conditions for the two experiments. As shown in Fig. 2.1, the experiment at Xingerxi Block Area was carried out under the well pattern of "four injection wells and nine production wells," and each injection well of the combinational system was adjacent to the peripheral water-injection well. During the process of slug injection, the high reservoir pressure

TABLE 2.6 Industrial test program data of combination flooding in No. 4 oil production plant in Daqing Oilfield.

Scheme	Well number		Producer-injector spacing (m)	Preslug		Combinational system slug (main)			Combinational system slug (auxiliary)			Polymer slug		Enhanced value of recovery efficiency (%)
	Injection	*Production*		*P concentration (mg/L)*	*Volume (PV)*	*S concentration (%)*	*P concentration (mg/L)*	*Volume (PV)*	*S concentration (%)*	*P concentration (mg/L)*	*Volume (PV)*	*Concentration (mg/L)*	*Volume (PV)*	
Middle Xingbei Block 2	17	27	250	1400	0.128	0.2	1000	0.354	0.2	1000	0.116	1150	0.2	18.05
Southern Block 5	29	39	175	1200	0.0617	0.2	1650	0.378	0.2	1650	0.048	1200	0.2	18.10
North Block 1 East Fault	49	63	125	1300	0.054	0.2	1900	0.429	0.2	1650	0.150	1000	0.2	18.42
North Block 3 West	13	14	250	800	0.018	0.1	1580	0.351	0.1	1580	0.104	1200	0.2	18.22

of the central area enclosed by four injection wells could be released towards the adjacent peripheral low-pressure area, which thereby ensures that the injection well can own a relatively high injection capacity. Under the condition of same injection volume, the injection fluid could have a higher polymer concentration. Fig. 2.4 depicts the sketch of well location of the “central area” of the well pattern in the industrial experiment. Each injection well of the combinational system in this area was not adjacent to the water-injection well; it was completely surrounded by the injection wells of the combinational system. The central area is a “closed” area. During the injection process of the combinational system, the pressure of stratum in this area rose relatively faster without pressure relief, and the liquid-injection capacity of the injection well significantly decreased. Under the condition of ensuring the same injection volume, it had to use the low-concentration slug injection.

As shown in Fig. 2.2, “one injection and one production” model containing $9 \times 9 \times 3$ nodes extended rightward to achieve the “four-injection and three-production” model containing $49 \times 9 \times 3$ nodes. All four injection wells of the central area were in the front row, including two “corner wells”; the three oil-production wells were in the internal row, which are all “side wells.” The “four-injection and three-production” model was used in the digital geology model established in the experiment at Xingerxi Block Area, and the experiment scheme of Middle Xingbei Block 2 and corresponding flooding scheme were carried out. The calculation results are detailed in Table 2.7. Fig. 2.5 depicts the oil-well water cut change curve (partial) and calculation

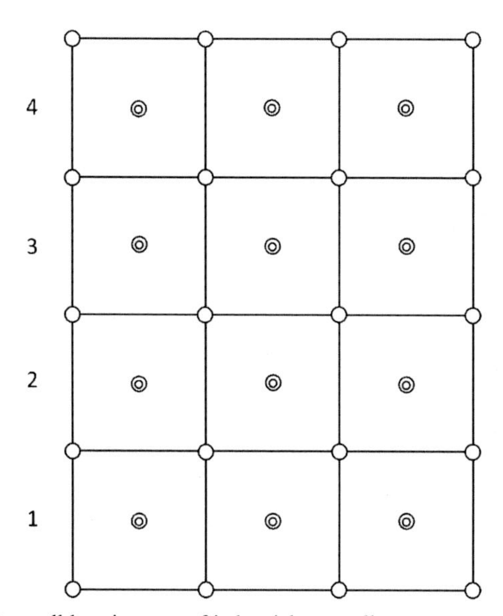

FIG. 2.4 The center well location map of industrial test well pattern.

TABLE 2.7 Digital test data of industrial combination flooding in No. 4 oil production plant in Daqing Oilfield.

Experiment	Flooding mode	Implementation time (t)	Maximum average pressure ratio of reservoir	Maximum viscosity of displacing phase (mPa s)	Recovery percent (%)	Enhanced recovery percentage (%)	Stratified residual oil (%)			S consumption (t)	P consumption (t)	Equivalent of total chemicals consumption P (t)	Oil increase for t equivalent polymer (t/t)
							Upper layer	*Middle layer*	*Lower layer*				
Middle Xingbei Block 2	Water flooding			0.6	47.27		46.08	41.20	35.59				
	Combination flooding	2746	1	14.42	65.18	17.91	39.14	25.46	16.53	802.1	750.3	1953.5	49.64
Southern Block 5	Water flooding			0.6	45.51		46.92	42.66	37.38				
	Combination flooding	962	0.8887	17.84	64.08	18.57	41.31	26.18	16.22	193.7	281.5	572.1	86.11
North Block 1 East Fault	Water flooding			0.6	43.72		48.53	43.94	38.68				
	Combination flooding	635	0.8968	23.12	69.70	25.98	37.13	19.70	14.12	223.9	257.7	593.6	59.23
North Block 3 West	Water flooding			0.6	42.36		50.56	44.51	39.24				
	Combination flooding	1865	0.7046	18.12	59.73	17.37	45.33	31.33	17.17	352.6	598.3	1127	83.44

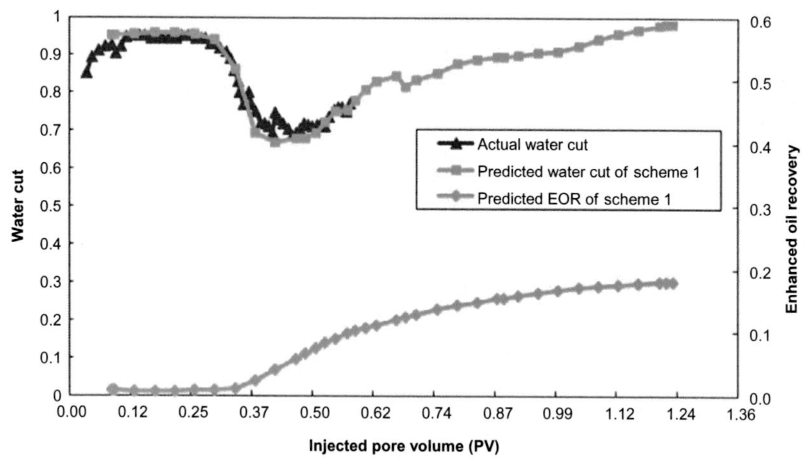

FIG. 2.5 Actual production curve and scheme matching curve of Xingerxi Block Area combination flooding.

curve of the central area. From the figure, we can learn that the two are in a good coincidence. The figure shows the change curve of enhanced recovery percentage. Upon the calculation, the enhanced oil recovery was 17.91%, close to 18.05% provided in the literature. It is specially noted that the maximum average pressure P2 of the oil reservoir from the calculation was basically equal to the maximum average pressure P1 obtained from the experiment at Xingerxi Block Area, with a difference of 0.9909; P2 is slightly less. The neighboring oil reservoirs in the same oil area shall have the same maximum average pressure. The high-precision coincidence of P1 and P2 indicate that the calculations of flooding scheme were correct and the models used in the calculations were credible and believable.

The "four-injection and three-production" calculation model was directly used to calculate and research the experiments of Southern Block 5 No. 2 Oil Production Plant, Northern Block 1 Duandong of No. 1 Oil Production Plant and North Block 3 West of No. 3 Oil Production Plant. The flooding experiment was digitized and the corresponding digital geology model was established. Table 2.7 shows the relevant experiment results and data of three oil recovery plants. Table 2.8 lists the relevant information data of the geology model of four industrial experiments. From the geologic data in Table 2.8, it can be seen that the reservoir heterogeneity gradually increased from Xingbei Block 2 in the south to North Block 3 West in the north, and combination flooding residual oil saturation S_{or}^{H} relating to combination flooding effect also changed regularly. Table 2.7 shows that the recovery percentage of water flooding gradually decreased from South to North and the combination flooding recovery efficiency of North Block 1 East Fault of No. 1 Oil Production Plant was greater than 7.56% provided in the literature and corresponded to the information of

TABLE 2.8 Geological model key parameters of industrial combination flooding in No. 4 oil production plant in Daqing Oilfield.

Scheme	Stratified permeability (10^{-3} μm^2)			Saturation of bound water (%)			Saturation of water flooding residual oil (%)			Saturation of combination flooding residual oil S_{or}^{H} (%)		
	Upper layer	*Middle layer*	*Lower layer*	*Upper layer*	*Middle layer*	*Lower layer*	*Upper layer*	*Middle layer*	*Lower layer*	*Upper layer*	*Middle layer*	*Lower layer*
Xingbei Block 2	100	215	525	24.0	22.0	21.0	36.5	34.5	32.5	15.50	14.50	13.50
Southern Block 5	100	230	600	24.0	22.0	21.0	37.0	36.0	34.5	15.50	14.45	13.35
North Block 1 East Fault	100	250	750	24.0	22.0	21.0	38.0	37.0	36.0	15.50	14.40	13.20
North Block 3 West	100	275	825	24.0	22.0	21.0	38.5	38.0	37.0	15.50	14.35	13.05

good field flooding effect. The enhanced oil recovery, given value by references, and calculated results of the experiments of the remaining plants were similar, at 18% or so.

Two data in Table 2.7 should be explained. The first is that the experiments of Xingbei Block 2 and North Block 3 West were all implemented under the condition of producer-injector spacing of 250 m, and the implementation times were all "over the standard." Because the complete oil-well water cut change curve was tested, it was difficult to determine whether there is a surfactant failure during the flooding process. Second is that the 30% retention rate of system viscosity for the fitting calculation of experiments of South Block 2, North Block 1 East Fault and North Block 3 West was taken to calculate the reservoir maximum average pressure during the flooding process. Table 2.7 shows the comparative value between them and the corresponding pressure in the experiment at Middle Xingbei Block 2. In the subsequent calculations, the surfactant stability was defined by the safe implementation time determined by the experiment of Xingerxi Block Area, and the retention rate of system viscosity took 30%. The maximum average pressure value of the oil reservoir was taken as the injection pressure value of the flooding experiment of this oil reservoir.

2.3.2 Research of optimization flooding scheme of industrial experiment in No. 4 oil production plant in Daqing Oilfield

For this experiment we adopted the digital geology model established by fitting flooding experiments, the "four-injection and three-production" structural model, the reservoir injection pressure limit value determined by fitting experiment, and the technical indicators of the optimized flooding scheme acquired from the research results in the previous section. Two-stage structure optimization slug: the front stage was the combinational system slug. Given that surfactant concentration was 0.3%, slug volume was 0.3 PV; the rear stage was polymer protective slug. Given that the slug was with a volume of 0.6 PV and producer-injector spacing of 125 m and injection speed of 0.40 PV/y. The results are shown in Table 2.9. From the table, it can be seen that the implementation of all schemes did not exceed 1570 days, meeting the requirements of the experiment scheme. The maximum average pressures of the test oil reservoir were all lower than the safe pressure limit value, meeting the requirements of implementation time. The maximum average pressures of the test oil reservoirs were all lower than the corresponding safe pressure limit value and the EOR enhanced recovery percentages were all greater than 28%. The enhanced recovery percentage of North Block 1 West Fault of No. 1 Oil Production Plant was the highest, reaching 30%. The oil increase for ton equivalent polymer was greater than 60 tons.

It was clearly understood that it would be very difficult to produce the residual oil left under the ground. Various oil recovery plants at Daqing Oilfield would try the best to obtain the high recovery efficiency of crude oil by taking

TABLE 2.9 Optimization scheme results of industrial combination flooding in No. 4 oil production plant in Daqing Oilfield (well spacing is 125 m).

Experiment	Flooding mode	Implementation time (d)	Reservoir maximum average pressure ratio	Slug polymer concentration (mg/L)	Maximum viscosity of displacing phase (mPa s)	Recovery percent (%)	Enhanced recovery percentage (%)	Stratified residual oil (%)			S consumption (t)	Equivalent of total chemicals consumption P (t)	Oil increase for ton equivalent polymer (t/t)
								Upper layer	*Middle layer*	*Lower layer*			
Middle Xingbei Block 2	Water flooding				0.6	47.39		45.55	41.22	35.82			
	Combination flooding	1052	0.9982	2030	27.26	75.66	28.27	26.15	16.40	14.16	174.1	614.6	60.47
Southern Block 5	Water flooding				0.6	45.55		46.69	42.68	37.50			
	Combination flooding	1045	0.8832	1999	26.48	74.15	28.60	29.47	16.70	14.08	174.1	617.7	60.87
North Block 1 East Fault	Water flooding				0.6	43.73		48.51	43.93	38.67			
	Combination flooding	1029	0.8946	2185	30.44	74.04	30.31	30.44	16.38	13.68	174.1	641.6	62.11
North Block 3 West	Water flooding				0.6	42.39		49.90	44.76	39.56			
	Combination flooding	1013	0.7036	1950	25.62	71.35	28.96	35.61	17.43	13.77	174.1	600.7	63.34

advantage of the corresponding optimized flooding schemes in conjunction with the oilfield situations so as to ensure as little crude oil as possible is left underground.

2.4 Conclusions

(1) The software that describes the combination flooding mechanism and the simplified model that describes the main reservoir geological features and relevant combination flooding information were used for the fitting calculation of the field flooding experiment. The flooding experiment was digitized and the digital geologic model platform was established, on which the flooding scheme was calculated.

(2) The digital research conducted at the Xingerxi Block Area achieved important results. It confirmed that the enhanced oil recovery of the corresponding water flooding for the combination flooding experiment increased to 25% or so, indicating a successful experiment. The ultra-low interfacial tension and high-viscosity system was researched and confirmed, which ensured a high enhanced oil recovery for the flooding experiment. Through the experiments, it is understood that using the appropriate combinational system slug and the high-viscous, long-volume polymer slug to further improve the recovery percent of low-permeability layer crude oil.

(3) We established the digital geology model of combination flooding on the basis of the successful combination flooding experiment and determined the key technical parameters of the scheme. We studied the impact of factors such as well spacing, injection speed, system viscosity, interfacial tension, and others on the flooding effect according to the technical principle of capillary number curve. We designed and optimized the flooding scheme, and based on the corresponding digital geology model established according to the combination flooding industrial field experiments of No. 4, No. 2, No. 1, and No. 3 Oil Production Plants, we carried out and optimized the flooding scheme under the condition of producer-injector spacing of 125m. The corresponding water flooding EORs of four oil production plants can all be greater than 28%. The highest enhanced recovery percentage of No. 1 Oil Production Plant can reach 30% and the oil increase for ton equivalent polymer is greater than 60 tons.

(4) The high-level research results confirm that the capillary number experimental curve *QL* describes more completely and accurately the corresponding relationship of capillary number and residual oil saturation during the combination flooding process. The use of "capillary number experiment curve *QL*" and "phase permeability curve *QL*" in the IMCFS software describes more accurately the flooding mechanism of combination flooding and the oil-water relative motion law during the combination flooding process. It is necessary to significantly enhance the numerical simulation research accuracy of chemical flooding, calculate and obtain the richer and more applicable information for the establishment of combination flooding digital geology model

platform so as to make the calculation study of chemical flooding scheme to be stepped from "qualitative research" into "quantitative research." Therefore, the "numerical simulation research" of chemical flooding field experiment can be called "digital flooding experiment research." This change in title sets a higher requirement on research work.

Symbol descriptions

V	seepage velocity of displacement phase, m/s
μ_w	viscosity of displacing phase, mPa s
σ_{ow}	interfacial tension between displacing phase and displaced phase, mN/m
N_c	capillary number value, dimensionless
N_{ct1}	limit capillary number when conversion of driving condition occurs during combination flooding process, dimensionless
N_{ct2}	limit capillary number when the residual oil value in combination flooding does not decrease any longer under "Type I" driving condition, dimensionless
S_{or}^H	lowest residual oil saturation in combination flooding process under "Type I" driving condition, i.e., residual oil saturation corresponding to capillary number lies between limit capillary number N_{ct2} and N_{ct1}
"Type I"	driving status: displacement under the condition that the capillary number is less than or equal to the limit capillary number N_{ct1}
"Type II"	driving status: displacement under the condition that the capillary number is higher than the limit capillary number N_{ct1}

References

1. Meyers JJ, Pitts MJ, Wyatt K. *Alkaline-Surfactant-Polymer Flood of the West Kiehl, Minnelusa Unit*. SPE 24144, 1992:423–435.
2. Vargo J, Turner J, et al. *Alkaline-Surfactant-Polymer Flooding of the Cambridge Minnelusa Field*. SPE 55633, 1999:1–6.
3. Felber BJ. *Selected U.S. Department of Energy's EOR Technology Applications*. SPE 84904, 2003:1–11.
4. Wang F, Wu X, Chen G, et al. Technical progress of alkaline-surfactant-polymer flooding (ASP) in Daqing Oilfield. *Petrol Geol Oilfield Dev Daqing*. 2009;28(5):154–162.
5. Qi LQ, Liu ZZ, Yang CZ, et al. Supplement and optimization of classical capillary number experimental curve for enhanced oil recovery by combination flooding. *Sci China Technol Sci*. 2014;57:2190–2203.
6. Wang D, Cheng J, Wu J, et al. *Summary of ASP Pilots in Daqing Oil Field*. SPE 57288, 1999.
7. Qi L. *Numerical Simulation Research of Polymer Flooding Engineering*. Beijing: Petroleum Industry Press; 1998.
8. Li S, Zhu Y, Zhao Y, et al. Evaluation of pilot results of alkali-surfactant-polymer flooding in Daqing Oilfield. *Acta Petrol Sin*. 2005;26(3):56–63.

Chapter 3

Research on laboratory combination flooding experiment digitization

Lianqing Qi[a,b], Zhongbin Ye[c], Yanjun Yin[b], Chengsheng Wang[b], Chuntian Liu[a], Guanli Xu[d], Shijia Chen[b], and Feng Li[b]

[a]*CNPC Daqing Oilfield Exploration and Development Research Institute, Daqing, China,* [b]*CNOOC Energy Technology & Services-Drilling & Production Technology Services Co., Tianjin, China,* [c]*School of Chemistry and Chemical Engineering of Southwest Petroleum University, Chengdu, China,* [d]*SINOPEC Petroleum Exploration and Production Research Institute, Beijing, China*

Chapter outline

Development and Application of Classical Capillary Number Curve Theory
https://doi.org/10.1016/B978-0-12-821225-7.00003-5

Indoor flooding experiment is a research method dominated by alkali-surfactant-polymer (ASP) flooding. The flooding tests on the physical model are usually used to understand the flooding mechanism and optimize the flooding scheme. The flooding experiments are completed mainly on a "2D cross-section model," which is an artificial three-layer vertical heterogeneous rectangular core. Because the flooding experiments in the "2D cross-section model" cannot reflect the displacing effect of two sides of the mainstream line of the underground reservoir, the flooding experiment can be used only for qualitative comparative studies. In Chapter 2, the authors established a digital flooding experiment platform via the fitting calculation research of combination flooding field experiments. Using this platform, they carried out an in-depth study of combination flooding technology via digital flooding experiments. However, this method was carried out under the condition of existing field experiments. What we need is to research and establish the corresponding digital flooding experiment platform on the basis of flooding experiments in combination with oil reservoir conditions, and then use it to research the digital flooding experiments.

3.1 Discussion on conditions of "equivalent" relationship between indoor flooding experiments and field experiments under the condition of chemical flooding

In order to establish the condition of "equivalent" relationship between indoor flooding experiments and field flooding experiments, the model design and the injection velocity design of fitting calculation are given below.

3.1.1 Design of equivalent research model

Geology models shall be designed according to the research needs and be simplified as much as possible on the basis of ensuring compliance with the scientific principle and clearly interpreting problems. Based on this ideology and according to the combination of heterogeneous core structure analysis with simulation calculation, Qi[1] proposed the geology model design of different heterogeneous reservoir simplified structures: oil reservoir plane homogeneity and vertical heterogeneity three-layer structure. Reservoir heterogeneity coefficient of variation (V_K) and corresponding interval permeability are listed in Table 3.1. Different permeability layers were permuted and combined into the heterogeneous oil reservoirs of different rhythmic sedimentations.

Because of the plane homogeneity of the oil reservoir, the simulation calculation studies could be carried out on the model of two wells in one-fourth well group in a well pattern of five-point methods, including one injection well and one production well. Fig. 3.1 shows that 9×9 grids could be taken in the plane and oil/water wells were separated by eight grids. This figure displays the concentration change of well grid chemical substances and the change of their physical-chemical properties as well as displays the effect of chemical agents during the flooding process.

TABLE 3.1 Distribution of vertical permeability under different V_K.

V_K	0	0.248	0.433	0.590	0.720	0.820	0.890	0.968
K_1 μm^2	0.442	0.0987	0.0987	0.0987	0.0987	0.0987	0.0987	0.0987
K_2 μm^2	0.442	0.1234	0.1579	0.2073	0.2961	0.4935	0.7403	1.974
K_3 μm^2	0.442	0.1974	0.3158	0.5182	0.8883	1.4085	2.2208	5.922

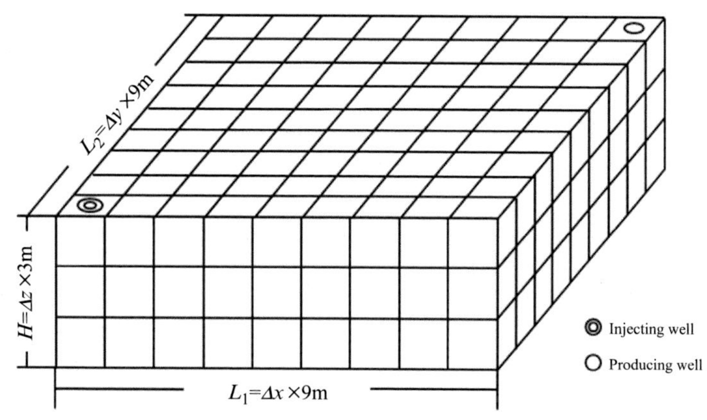

FIG. 3.1 Geology model structure.

Based on this model design, the authors used numerical simulation methods to research polymer flooding technology. The results (see Chapter 2) withstood the tests of polymer flooding practices at Daqing Oilfield.

This model design was continuously used in the numerical simulation calculation studies of combination flooding. In Chapter 2, the authors used the simplified model to conduct calculation research of the combination flooding field experiments [2,3] at Xingerxi Block Area. Fig. 3.2 plots the oil-well water-cut change curve of the field experiment, enhanced oil recovery (EOR) percentage change curve, and corresponding curve acquired via the fitting calculation. The comparative studies indicate that the calculation results had a good enough accuracy.

The core of flooding experiments used the same heterogeneity coefficient of variation V_K and stratified permeability as the simplified structure geology model. According to the indoor test conditions, it was recommended to use the artificial core of plane homogeneity and vertical three-layer equal thickness. The plane side length was 30–40 cm and total thickness of the three layers was 3–5 cm.

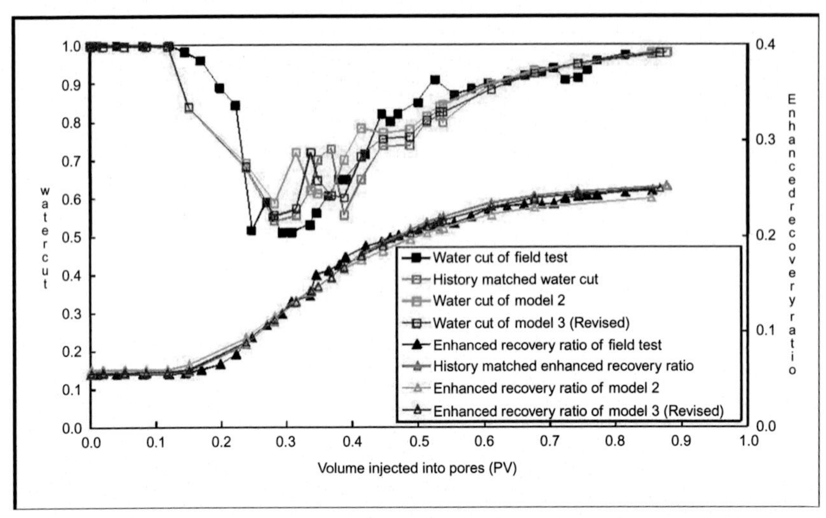

FIG. 3.2 Result curve of Xingerxi Block Area field combination flooding experiment and plan.

3.1.2 Design of injection velocity

The simulation calculation studies confirmed that the displacing velocity was technically crucial to the "equivalent" researches of combination flooding field experiments and indoor core flooding experiments.

Under the reservoir conditions of variation coefficient of 0.59 and total thickness of 12m, we researched the flooding effect of water flooding and combination flooding schemes under different well spacing. The related comparison schemes were calculated by changing the reservoir thickness according to the research needs.

Table 3.2 lists the results of the water flooding scheme, from which we learn that the change of well spacing exerted almost no impact on flooding effect.

After this, we researched combination flooding using united injection velocity of 0.15PV and well spacing of 250m. The combinational system slug was 0.3PV, the subsequent polymer slug was 0.7PV, the combinational system interfacial tension of former group scheme was 0.005mN/m, the interfacial tension of latter group scheme was 0.00125mN/m, and the two-stage slug polymer concentrations of the schemes were all 2000mg/L. The calculation results are listed in Table 3.3.

The result of scheme group 1 was analyzed when the system interfacial tension was 0.005mN/m. It is clear that the maximum underground viscosity of the system decreased with the reduction of well spacing, because the seepage velocity increased the shearing effect with the reduction of well spacing. It is noted that the change of maximum grid capillary number during the flooding process (about four times the minimum capillary number in the scheme) indicates the influence of the seepage velocity. The average seepage velocity on the

TABLE 3.2 Result of water flooding under different well spacing.

Well spacing (m)	Total thickness (m)	Recovery efficiency (%)	Stratified residual oil (%)		
			Upper layer	*Middle layer*	*Lower layer*
250	12	47.16	46.23	41.30	35.58
176	12	47.20	45.95	41.36	35.72
150	12	47.22	45.84	41.37	35.77
125	12	47.24	45.75	41.37	35.82
	6	47.20	46.05	41.33	35.66
	3	47.18	46.20	41.30	35.56
88	12	47.29	45.58	41.33	35.90
	6	47.21	45.86	41.37	35.76
	3	47.21	46.07	41.31	35.63
62.5	12	47.38	45.42	41.24	35.94
	6	47.24	45.70	41.38	35.85
	3	47.21	45.96	41.34	35.70
54.47	12	47.42	45.41	41.18	35.93

"cross-section" of the oil reservoir (spanning the middle point of the mainstream line and perpendicular to the mainstream line) was calculated and studied. It was found that there was a significant difference in seepage velocity under different well spacing. The difference of flooding effect was then researched. The results show that the combination flooding recovery efficiency significantly improved and EOR gradually increased with the reduction of well spacing. The results for the latter group scheme with system interfacial tension of 0.00125 mN/m show that the average seepage velocity of the cross-section was the same as in the former scheme and the system's viscosity changes were similar. Due to the significant reduction of system interfacial tension, the capillary number of underground grid during the oil flooding process increased proportionally. Attention should be paid to the fact that combination flooding recovery efficiency and the EOR synchronously increased with the reduction of well spacing in this group scheme. The research shows that the performance of combination flooding is affected by many factors, such as changes in well spacing. Only the injection velocity was changed for the test in the same flooding scheme—the same average seepage velocity of the cross-section was maintained under different well spacing. The calculation results are listed in Table 3.4.

TABLE 3.3 Result of compound flooding in different well spacing under the condition of Daqing injection velocity.

Interfacial tension (mN/m)	Well spacing (m)	Average seepage velocity of cross section (cm/d)	Max. underground system viscosity (mPa s)	Max. underground grid capillary number during the flooding	Stratified residual oil (%)			Combination flooding recovery (%)	EOR (%)
					Upper layer	*Middle layer*	*Lower layer*		
0.005	250	5.77	24.60	0.0223	31.91	21.17	15.18	70.70	23.54
	176	8.20	24.45	0.0310	30.44	19.07	14.74	72.43	25.23
	150	9.62	24.12	0.0363	29.72	18.46	14.58	73.07	25.85
	125	11.55	23.90	0.0431	28.96	18.03	14.44	73.64	26.40
	88	16.40	22.94	0.0599	27.97	17.65	14.31	74.28	26.99
	62.5	23.10	22.39	0.0830	27.51	17.56	14.28	74.53	27.15
	54.47	26.51	22.18	0.0943	27.48	17.63	14.29	74.51	27.09
0.00125	250	5.77	24.38	0.0897	24.99	16.93	14.27	75.89	28.73
	176	8.20	23.82	0.1238	25.68	16.72	14.25	75.69	28.49
	150	9.62	23.66	0.1444	25.91	16.65	14.24	75.62	28.40
	125	11.55	23.20	0.1732	26.10	16.62	14.21	75.57	28.33
	88	16.40	22.72	0.2396	26.29	16.74	14.21	75.44	28.15
	62.5	23.10	22.29	0.3303	26.46	16.93	14.08	75.34	27.96
	54.47	26.51	22.16	0.3817	26.72	17.08	13.97	75.21	27.79

TABLE 3.4 Result of compound flooding in different well spacing while the same average seepage velocity in cross-section.

Interfacial tension (mN/m)	Well spacing (m)	Average Seepage velocity of cross section (cm/d)	Max. underground system viscosity (mPa s)	Max. underground grid capillary number during the flooding	Stratified residual oil (%)			Combination flooding recovery (%)	EOR (%)
					Upper layer	*Middle layer*	*Lower layer*		
0.005	250	5.77	24.60	0.0223	31.91	21.17	15.18	70.70	23.54
	176	5.77	24.21	0.0222	32.58	21.47	15.16	70.30	23.10
	150	5.77	24.14	0.0222	32.79	21.56	15.15	70.18	22.96
	125	5.77	24.26	0.0222	32.88	22.06	15.12	69.93	22.69
	88	5.77	24.43	0.0221	33.31	22.90	15.14	69.38	22.09
	62.5	5.77	23.68	0.0221	33.60	23.63	15.21	68.91	21.53
	54.47	5.77	24.05	0.0220	33.75	23.83	15.27	68.74	21.32
0.00125	250	5.77	24.38	0.0897	24.99	16.93	14.27	75.89	28.73
	176	5.77	24.23	0.0889	26.58	16.92	14.27	75.21	28.01
	150	5.77	24.17	0.0891	27.20	16.95	14.27	74.93	27.71
	125	5.77	24.23	0.0891	27.67	16.96	14.27	74.72	27.48
	88	5.77	24.32	0.0885	28.47	17.20	14.28	74.28	26.99
	62.5	5.77	23.64	0.0883	29.30	17.69	14.31	73.69	26.31
	54.47	5.77	23.80	0.0879	29.86	17.94	14.34	73.33	25.91

The conditions of the two group schemes were basically the same. The system viscosity was slightly less under the condition of relatively narrow well spacing, and the maximum grid capillary number during the flooding process became somewhat smaller with the reduction of well spacing. The combination flooding recovery and the EOR of the schemes decreased with the reduction of well spacing, with a maximum difference of less than 3%. The research outcomes of stratified residual oil changes indicate that the residual oil value of the upper low-permeability layer significantly increased from the top down in the same group scheme. The residual oil value of the middle middle-permeability layer slightly increased. The residual oil value of the lower high-permeability layer was nearly identical. Through comparison of the two groups of schemes, the change of the group with lower interfacial tension was significant. It is clear that the fingering phenomenon of combination flooding is relatively serious with the reduction of well spacing; this is the basic feature of combination flooding.

The impact of the thickness change of the oil reservoir on the flooding effect was researched for three kinds of well spacing conditions at random. We used well spacing of 125 m, 88 m, and 62.5 m, respectively, and total thickness of oil reservoir of 12 m, 6 m, and 3 m, respectively. Under the condition of maintaining the equal average seepage velocity of the cross-section, the corresponding schemes were calculated; the results are listed in Table 3.5. Under the condition of same well spacing and same interfacial tension system, the oil reservoir thickness decreased and the underground viscosity and the corresponding capillary number tended to do the same. The analysis of the recovery effect of the schemes indicated that the oil reservoir thickness decreased while the combination flooding recovery efficiency and EOR increased. The studies on the change of stratified residual oil suggested that the residual oil value was closer for both schemes in the middle high-permeability layer. However, in the low-permeability layer, the residual oil value of the thin oil reservoir scheme was relatively lower. The studies show that the negative impact of poor flooding effect resulted from the reduction of well spacing would be offset by increased flooding effect when the oil reservoir thickness appropriately thinned. From the table, we can see that under two kinds of system interfacial tensions, the flooding effect of flooding scheme for well spacing of 62.5 m and oil reservoir thickness of 3–6 m was close to that of flooding scheme for well spacing of 125 m and oil reservoir thickness of 6–12 m.

In this section, we discussed the impact of injection velocity on flooding effect under the change conditions of well spacing and oil reservoir thickness. The 3D flooding experiment model could be regarded as a miniature field experiment, in which there is an extreme producer-injector well spacing and relatively thin oil reservoir thickness. The reduction of producer-injector well spacing from the oil reservoir well spacing caused a relatively poorer flooding effect, while the reduction of oil reservoir thickness improved the flooding effect. The offset between them would likely ensure the "equivalent" fitting calculation to reach a relatively satisfactory outcome.

TABLE 3.5 Result of compound flooding in different well spacing and thickness while the same average seepage velocity in cross-section.

Interfacial tension (mN/m)	Well spacing (m)	Total reservoir thickness (m)	Average seepage velocity of cross section (cm/d)	Max. underground system viscosity (mPa s)	Max. underground grid capillary number during the flooding	Stratified residual oil (%)			Combination flooding recovery (%)	EOR (%)
						Upper layer	*Middle layer*	*Lower layer*		
0.005	125	12	5.77	24.26	0.0222	32.88	22.06	15.12	69.93	22.69
		6	5.77	24.25	0.0223	32.31	21.82	15.11	70.29	23.09
		3	5.77	24.47	0.0224	31.48	21.17	14.94	71.00	23.82
	88	12	5.77	24.41	0.0221	33.31	22.90	15.14	69.38	22.09
		6	5.77	24.05	0.0221	32.77	22.18	15.19	69.90	22.69
		3	5.77	24.22	0.0222	32.24	21.60	15.03	70.45	23.24
	62.5	12	5.77	23.68	0.0221	33.60	23.63	15.21	68.91	21.53
		6	5.77	23.62	0.0221	32.94	22.89	15.28	69.49	22.25
		3	5.77	23.82	0.0222	32.74	22.01	15.13	70.01	22.80
0.00125	125	12	5.77	24.23	0.0891	27.67	16.96	14.27	74.72	27.48
		6	5.77	24.26	0.0894	25.14	17.17	14.26	75.73	28.53
		3	5.77	25.11	0.0897	22.91	17.04	14.19	76.77	29.59
	88	12	5.77	24.32	0.0885	28.46	17.20	14.28	74.28	26.99
		6	5.77	24.09	0.0888	26.30	17.31	14.30	75.15	27.94
		3	5.77	24.65	0.0890	24.13	17.20	14.23	76.15	28.94
	62.5	12	5.77	23.64	0.0883	29.30	17.69	14.31	73.69	26.31
		6	5.77	23.55	0.0884	27.28	17.59	14.34	74.59	27.35
		3	5.77	24.02	0.0888	25.51	17.61	14.29	75.36	28.15

3.2 Research of indoor flooding experiment

Table 3.6 lists the results of one group of flooding experiments completed in 2000. The 3D flooding experiment core was developed on the basis of researching the oil reservoir of Xingerxi Block Area at Daqing Oilfield. The geometric dimensions of the reservoir were $32 \times 32 \times 3.6$ cm. The core was subject to plane homogeneity and vertical heterogeneity positive rhythm, with an equal thickness for three intervals, and the heterogeneous coefficient of variation of the model was about 0.59. According to the core fabrication needs, the permeability ratio of various intervals was determined as $0.2\,\mu m^2$, $0.6\,\mu m^2$, and $1.2\,\mu m^2$, respectively.

For the oil displacement experiment, it was transferred to the combination flooding after the water content in produced fluid reached 98% in the water flooding process. The flooding process would be terminated when the water content in produced fluid decreased and then increased to 98% again. The combinational system slug volume of combination flooding scheme was 0.3 PV. The two-stage polymer slug was adopted in the subsequent experiment. The former belonged to the sewage preparation volume of 0.2 PV and the latter belonged to freshwater preparation volume of 0.3 PV; the injection velocity was 0.6 mL/min.

The data shows a group of good test results. The difference between the maximum volume and the minimum volume of saturated water was 50 mL, thereby determining that the biggest difference of core porosity was 1.3%. The difference between the maximum and the minimum of oil saturation was 2%, indicating that the technical level for both core fabrication and saturated oil/water operations during the experiment was relatively higher. The lowest water flooding recovery of the test data was 46.4% and the highest was 50.0%, with the biggest difference being 3.6%. This is a relatively ideal test result.

3.3 Fitting calculation of flooding effect of indoor flooding experiment based on flooding scheme of field geology model

The combination flooding software IMCFS[4] was used for the fitting calculation. The software described the flooding mechanism of combination flooding and oil-water relative movement law during the flooding process of combination flooding with "capillary number experiment curve *QL*" and "phase permeability curve *QL*." It also depicted the oil reservoir by means of separating the "tanks." The same "tank" area has the same oil reservoir characteristic data, especially the data of its own capillary number curve and phase permeability curve of water flooding and combination flooding.

The geology model was established under the geologic reservoir condition of Xingerxi Block Area. The core porosity data in Table 3.6 were analyzed, and the average porosity of eight cores was found to be 0.26. According to the

TABLE 3.6 Result of ASP flooding experiment in different interfacial tensions.

Experiment	Test core data			ASP system parameter			Polymer slug		Test result			
Core No.	*Saturated water volume (mL)*	*Oil saturation (%)*	*Porosity (%)*	*Place of origin for surfactant*	*Interfacial tension 10^{-2} (mN/m)*	*System viscosity (mPa s)*	*Sewage preparation viscosity (mPa s)*	*Freshwater preparation viscosity (mPa s)*	*Max. injection pressure (10^5Pa)*	*Water flooding recovery (%)*	*Combination flooding recovery (%)*	*EOR (%)*
AS-8	921	74.8	25.0	Daqing	4.08[a]	19.2	20.5	19.5	2.942	47.2	71.9	24.7
AS-3	959	74.9	26.0	US	2.11	21.4	18.7	20.6	2.804	50.0	74.8	24.8
AS-2	966	75.5	26.2	Daqing	1.30	20.7	20.0	20.6	2.929	49.5	74.2	24.7
AS-15	971	73.8	26.3	Beijing	0.38[a]	18.9	20.5	19.7	1.860	46.9	70.8	23.9
AS-7	946	74.2	25.7	Dalian	0.34	21.1	20.0	20.4	1.397	48.4	73.5	25.1
AS-4	968	75.2	26.3	Dalian	0.31	19.5	20.0	20.6	2.311	49.9	74.2	24.3
AS-1	969	74.3	26.3	Beijing	0.21	20.5	17.8	20.6	1.819	48.7	76.6	27.9
AS-14	959	73.5	26.0	Daqing	0.078	22.5	20.6	19.3	1.245	46.4	70.3	23.9

[a]*The weak base system and others are the strong base system.*

calculation of core dimension and injection velocity of flooding experiment, the average seepage velocity of the core cross-section was 2.36×10^{-4} cm/s. The model in Fig. 3.1 was taken for the simulation calculation under the field oil reservoir condition, with a grid of $9 \times 9 \times 3$, total reservoir thickness of 12 m, porosity of 0.3, and an injection velocity of 58.48 m^3 as is usually used at Daqing Oilfield. The results show that the average seepage velocity of the geology model cross-section was 2.37×10^{-4} cm/s under the condition that the horizontal grid spacing was 6.25 m, which meets the "equivalent fitting" requirement. The average seepage velocity of the cross-section for both flooding experiment model and fitting calculation geology model was equal. The heterogeneity variation coefficient of the oil reservoir at Xingerxi Block Area was approximately 0.59. As is shown in Table 3.1, the permeability of three intervals was 0.0987 μm^2, 0.2073 μm^2, and 0.5182 μm^2, respectively. It was confirmed through calculation that the water flooding recovery efficiency was 47.2%, which is the specified heterogeneity variation coefficient and water flooding recovery efficiency of the target oil reservoir.

The data in Table 3.7 shows that the water flooding recovery efficiency was within 46.4%–50.0%, which resulted from the existing differences between the model fabrication and interlayer heterogeneity. The water flooding scheme was calculated by using the geology model with a heterogeneity variation coefficient of 0.433, and the recovery efficiency thus acquired was 48.02%. The water flooding recovery efficiency of five experiments among eight was higher than 48.02%. Therefore, the corresponding heterogeneity variation coefficient of the five experiments was less than 0.433, far less than the heterogeneity variation coefficient of 0.59 at Xingerxi Block Area confirmed by the simulation calculation. Hence none of them should be listed in the selected targets.

On the basis of summarizing the fitting research experiences of the flooding experiment at Xingerxi Block Area, the fractional fitting calculation and research method was established.

First we fit or matched the process of the water flooding experiment and trimmed the relevant data of stratified permeability of oil reservoir and phase permeability curve—stratify water flooding residual oil saturation, fit or match the water flooding recovery indicator, determine the basic physical parameters of oil reservoir. On the basis of satisfactory water flooding fitting results, we fit or matched the experiment process of combination flooding, trimmed the relevant limit capillary number and combination flooding phase-permeability curve parameters—stratify the combination flooding residual oil saturation S_{or}^{H}, fit or match the combination flooding recovery indicators and determine the basic physical parameters of oil reservoir associated with the combination flooding.

The results of the fitting calculation are shown in Table 3.7. Through comparison with the data in Table 3.6, we can see that the recovery efficiency and the EOR of the two fitting calculations were nearly the same. Fig. 3.3 plots the actual and fitting calculation curves of the water-cut and EOR percentage for

TABLE 3.7 Calculation results of matching different flooding experiment.

Experiment No.	Interfacial tension 10^{-2} (mN/m)	Maximum underground viscosity (mPa s)	Injection pressure specific value	Max. grid capillary number during calculation process	Capillary number N_{ct2} Parameter in fitting calculation	Stratified residual oil (%)			Water flooding recovery efficiency (%)	Combination flooding recovery efficiency (%)	Enhanced oil recovery EOR (%)
						Upper layer	*Middle layer*	*Lower layer*			
AS-8	4.08	23.04	1.0059	0.0093	0.0002	31.26	18.14	13.81	47.2	71.91	24.71
AS-15	0.38	20.24	1	0.0841	0.0025	30.92	19.09	14.82	46.91	70.80	23.89
AS-14	0.078	20.28	0.8762	0.4392	0.0025	31.46	19.57	14.91	46.4	70.30	23.90
AS-14	**0.078**	**20.95**	**0.8563**	**0.3123**	**0.0025**	**31.80**	**18.99**	**15.13**	**46.4**	**70.31**	**23.91**

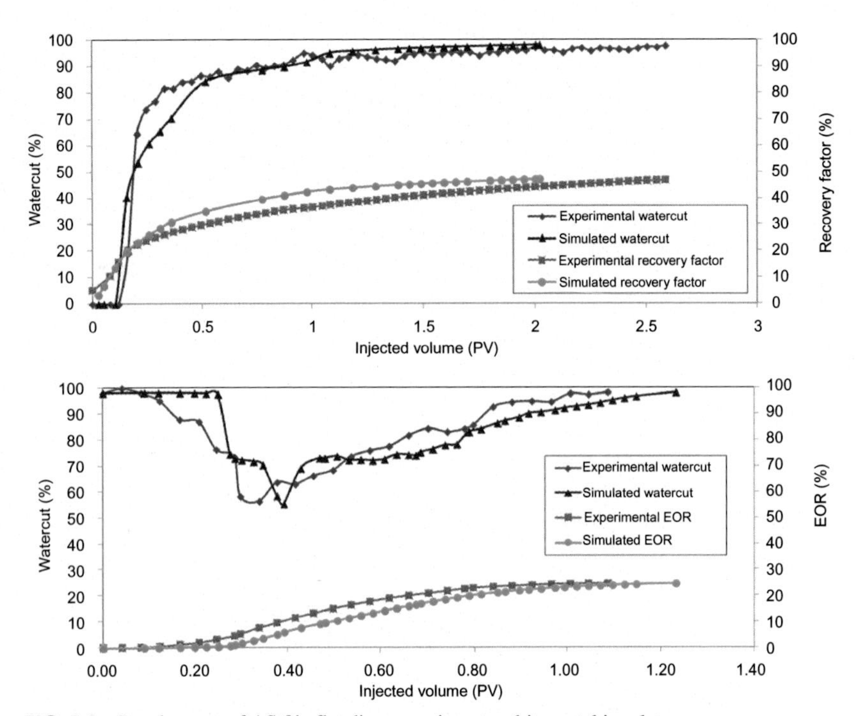

FIG. 3.3 Result curve of AS-8's flooding experiment and its matching data.

core AS-8 flooding experiments. The fitting calculation obtained a satisfactory result. Table 3.8 shows the related data of corresponding geology model water flooding and combination flooding of various experiments from previous fitting research.

The fitting calculations based on three experiments all achieved good results. By researching the fitting result of the AS-15 experiment, we acquired three-layer permeability of the geology model as 0.0987 μm^2, 0.2122 μm^2, and 0.5182 μm^2, respectively. These figures are very close to the specified stratified permeability data when the heterogeneity variation coefficient was 0.59. The fitted water flooding recovery was 46.91% and the experiment result was 46.9%. The enhanced recovery percentage of fitted combination flood was 23.89% and the experiment result was 23.9%. It should be noted that the limit capillary number N_{ct2} for achieving the minimum residual oil saturation of combination flooding was 0.0025, which was identical to that acquired from the fitting calculation of the field experiment at Xingerxi Block Area. This numerical value was identical to the relevant value in capillary number experimental curve *QL*.[4]

Then, we researched the flooding experiment of core number AS-8. The water flooding experiment recovery efficiency was 47.2%, indicating that the core was ideally fabricated. The flooding system interfacial tension of the

TABLE 3.8 Related data of water flooding and compound flooding from history matching.

Core No.	Stratified permeability (μm^2)			Bound water saturation (%)			Water flooding residual oil saturation (%)			Limit capillary number N_{ct2} parameter	Combination flooding residual oil saturation (%)		
	Low	*Medium*	*High*	*Low*	*Medium*	*High*	*Low*	*Medium*	*High*		*Low*	*Medium*	*High*
AS-8	0.1046	0.2270	0.5922	27	25	23	35	33.5	31.5	0.0002	15	13.5	13
AS-8	**0.1046**	**0.2270**	**0.5922**	**27**	**25**	**23**	**35**	**33.5**	**31.5**	**0.0025**	**15**	**13.5**	**13**
AS-15	0.0987	0.2122	0.5182	28	26	24	34.5	33.25	32	0.0025	15.6	14.8	13.9
AS-14	0.1036	0.2221	0.6662	28	26	24	35.4	33	31.7	0.0025	16.0	15.5	15.0
AS-14	**0.1036**	**0.2221**	**0.6662**	**28**	**26**	**24**	**35.4**	**33**	**31.7**	**0.0025**	**15.6**	**14.9**	**14.0**

combination flooding was 4.08×10^{-2} mN/m, and the system viscosity was 20 mPa s or so. Moreover, it was also noted that the maximum injection pressure in the experiments was higher than 2.8×10^5 Pa, higher than the maximum injection pressure 1.0×10^5 Pa for the core AS-15 experiment. This indicates that the viscoelasticity played the role for the oil flooding of high interfacial tension (10^{-2} mN/m) and a high viscous flooding system in a relatively homogeneous oil reservoir under the condition of relatively high reservoir pressure gradient, allowing for a higher EOR percentage to be reached. The results of the fitting calculation were then analyzed to research the core AS-8 experiment. Its fitting curve was plotted in order to highlight its good fitting effect. The fitting calculation process was analyzed and a satisfactory fitting result was obtained from the water flooding fitting calculation. The fitting of combination flooding adopted the same method as the fitting of the core AS-15 experiment and the stratified combination flooding residual oil saturation and the relevant capillary number all adopted the similar data of the fitting of core AS-15 experiment, all of which failed to reach the fitting requirements. It was then considered that the viscoelasticity of a high interfacial tension and high viscous system under the condition of high pressure gradient played the role and relaxed the requirements of reaching the minimum residual oil saturation of combination flooding—limit capillary number N_{ct2} decreased from 0.0025 to 0.0002, which met the fitting requirements. This treatment met the requirement of this particular case, but it was not desirable to preserve the limit capillary number N_{ct2} acquired from the particular fitting case as the oil reservoir parameters for the combination flooding, and therefore, the oil reservoir model obtained from the experiment fitting of this group could not be selected and used directly.

It should also be specified that a group of 3D core flooding experiments including AS-8, AS-3, and AS-2 in Table 3.6 proved that the flooding system with interfacial tension of 10^{-2} mN/m could reach the equivalent flooding effect of the system with its interfacial tension of 10^{-3} mN/m. For this reason, it was inappropriate for Ref.[5] to introduce or promote the flooding operations by using a low-concentration surfactant combinational system based on the good flooding effect of the flooding system with interfacial tension of 10^{-2} mN/m.

The model acquired from the fitting of the core AS-14 experiment was then analyzed. From the experiment data, we learn that the interfacial tension of flooding system was 7.8×10^{-4} mN/m, the system viscosity was 20 mPa s or so, water flooding recovery efficiency was 46.4%, and the EOR of combination flooding was 23.9%. It was also noted that injection pressure was the lowest among eight experiments. The low injection pressure and low EOR percentage are the obvious features of water-phase fingering that appeared in the high-permeability layer; the analysis of fitting calculation results shows that the recovery targets of water flooding and combination flooding were all similar to the experiment results. It is noted that the corresponding fitting data of the core AS-15 experiment were all relatively high by comparing the stratified residual oil values in the fitting results. There was also a relatively low oil reservoir pressure during the flooding process. It could be judged that the water-phase fingering appeared under the displacing condition of the extra-ultra-low interfacial tension system enlarged and distorted the combination flooding stratified residual oil saturation of the model acquired from the fitting. Therefore this model is not advisable.

3.4 Determination of digital flooding geology model

3.4.1 Assessment of digital model

The model acquired from the fitting of the core AS-15 experiment is the first model obtained from the fitting of flooding experiment, and therefore it is necessary to assess its applicability and accuracy. Chapter 2 introduced that the oil reservoir of Xingerxi Block Area had obtained the digital geology model through the fitting of field experiments and experienced the tests of the calculations and researches, which created the comparative conditions of assessing the model established by the flooding experiments. In this chapter, the different combination flooding schemes would be used on the two models to assess and check the digital reservoir geology models established by the fitting of the core AS-15 experiment. The schemes of three grades of interfacial tensions selected for the assessment are as follows:

- Scheme No. 1—interfacial tension $= 4.08 \times 10^{-2}$ mN/m
- Scheme No. 2—interfacial tension $= 0.38 \times 10^{-2}$ mN/m
- Scheme No. 3—interfacial tension $= 0.078 \times 10^{-2}$ mN/m

The calculation results of the schemes are listed in Table 3.9, among which, Model No. 0 is the fitting model of field experiment, and Model No. 1 is the model established by the fitting of the core AS-15 experiment.

First, we analyzed the calculation results of the scheme on Model No. 0. Its water flooding recovery efficiency complied with the requirements. As for the combination flooding scheme, Scheme No. 1 of high interfacial tension had a relatively low enhanced recovery percentage, around 10%. The enhanced recovery percentage of the scheme significantly increased with the reduction of well spacing. It also increased with the decrease of system interfacial tension. When it was lower than 10^{-3} mN/m, EOR could generally be more than 25%.

The comparative analysis of the calculation results of Model No. 1 was carried out. Its water flooding recovery efficiency was slightly lower than that of corresponding scheme of Model No. 0, with a maximum deviation of 0.45% and relative error of less than 1%; the comparison results of combination flooding schemes indicated that the EOR efficiencies of corresponding schemes for same well spacing in the same system were very similar. The maximum deviation of the corresponding schemes of system No. 1 and system No. 2 was less than 1%, the maximum deviation of corresponding schemes of system No. 2 was 2.63%, and the deviation of all other schemes was less than 2%. Through further

TABLE 3.9 Flooding effect of flooding schemes on different digital geology models.

Digital geology model	Well spacing (m)	Water flooding recovery efficiency (%)	1# Scheme system (4.08×10^{-2} mN/m)		2# Scheme system (0.38×10^{-2} mN/m)		3# Scheme system (0.078×10^{-2} mN/m)	
			Combination flooding recovery (%)	*EOR (%)*	*Combination flooding recovery (%)*	*EOR (%)*	*Combination flooding recovery (%)*	*EOR (%)*
Fitting Reservoir Geology Model 0 for Xingerxi experiment	250	47.16	56.47	9.31	71.76	24.60	74.94	27.78
	176	47.20	57.64	10.44	72.59	25.39	74.59	27.39
	150	47.22	58.38	11.16	72.93	25.71	74.30	27.08
	125	47.24	58.49	11.66	73.00	25.76	74.49	27.25
	88	47.29	61.43	14.14	73.09	25.80	74.36	27.07
Fitting Reservoir Geology Model 1 for Core AS-15 experiment	250	46.72	56.32	9.60	68.69	21.97	73.55	26.83
	176	46.75	57.33	10.58	70.29	23.54	73.34	26.59
	150	46.78	57.90	11.12	70.84	24.06	73.33	26.55
	125	46.79	58.53	11.74	71.16	24.37	73.13	26.34
	88	46.84	60.09	13.25	71.57	24.73	73.30	26.46
Fitting Reservoir Geology Model 2 for Core AS-8 experiment	250	47.00	66.78	19.78	75.66	28.66	75.66	28.66
	176	47.03	69.40	22.37	75.15	28.12	75.05	28.02
	150	47.06	70.27	23.21	75.19	28.13	75.07	28.01
	125	47.08	70.84	23.76	74.94	27.86	74.84	27.76
	88	47.14	71.67	24.53	74.53	27.39	74.44	27.30

Fitting Reservoir Geology Model 2 (amended) for Core AS-8 experiment	250	47.00	55.96	8.03	70.89	23.89	74.33	27.33
	176	47.03	57.03	10.0	71.82	24.79	74.13	27.10
	150	47.06	57.72	10.66	72.17	25.11	74.05	26.99
	125	47.08	58.55	11.47	72.36	25.28	73.79	26.71
	88	47.14	60.59	13.45	72.61	25.62	73.65	26.51
Fitting Reservoir Geology Model 3 for Core AS-14 experiment	250	46.14	55.04	8.90	68.01	21.87	71.52	25.38
	176	46.19	55.92	9.73	68.72	22.53	71.17	24.98
	150	46.22	56.56	10.34	69.15	22.93	71.09	24.87
	125	46.25	57.25	11.00	69.35	23.10	70.97	24.72
	88	46.32	59.12	12.80	69.66	23.34	71.03	24.71
Fitting Reservoir Geology Model 3 (amended) for Core AS-14 experiment	250	46.14	55.76	9.62	69.11	22.97	72.65	26.51
	176	46.19	56.72	10.53	69.81	23.62	72.22	26.03
	150	46.22	57.37	11.15	70.24	24.02	72.15	25.93
	125	46.25	58.08	11.83	70.46	24.21	71.95	25.70
	88	46.32	60.04	13.72	70.78	24.46	71.69	25.37

assessment, it was determined that Model No. 1 should be taken as the candidate model.

3.4.2 Calculation review and modification of two models

In Section 3.3, Model 2 established by the fitting of the core AS-8 experiment and Model 3 established by the fitting of the core AS-14 experiment were discarded. In this section, two models are used to calculate the corresponding schemes and review the calculations. The calculation results are shown in Table 3.9.

The analyses of the calculation results of Model 2 show that its water flooding recovery efficiency was closer to the result of Model No. 0 than that of Model No.1, and the maximum deviation of the corresponding schemes was 0.17%. The comparison of calculation results of combination flooding scheme of system No.1 shows that the difference amount the EOR of corresponding schemes was nearly doubled, and the EOR of the schemes of system No. 2 and system No. 3 was also much higher than that of corresponding schemes. The reasons for such a difference are clear; because the critical capillary number N_{ct2} was adjusted down to 0.0002. However, it is noted that the water flooding result of this model was closer to that of Model No. 0 and the combination flooding stratified residual oil saturation approached the corresponding data result of Model No. 1. Amend Model No. 2 and restore the critical capillary number N_{ct2} to 0.0025, which is identical to that of Model No. 1, and calculate the corresponding scheme on the amended model. The scheme result was also very close to the result of Scheme No. 0, and the maximum deviation of the recovery percentage of corresponding schemes was 0.87%. There was only the Scheme No. 1 of well spacing of 250 m whose deviation of EOR recovery was 1.28%, and the absolute value of the maximum deviation of all other schemes was less than 0.71%. The high-accuracy results are based on the fact that the water flooding result of this model was closer to the oil reservoir condition. The assessment indicated that the amendment was reasonable and the residual oil saturation of oil reservoir combination flooding was definite, as was the critical capillary number, which should not be changed by the displacing conditions. According to the assessment results, the amended model could also be regarded as the candidate model.

The calculation result of Model No. 3 was then analyzed. Its water flooding recovery percentage was about 1% lower than that of Model No. 0, and much lower than that of Model No. 1. We noted that in the case of combination flooding the EOR of corresponding schemes was generally lower than the calculation results of corresponding schemes of Model No. 0 and Model No. 1; it was more obvious for scheme No. 3 (extra-ultra-low interfacial tension). Thus, the impact of high residual oil saturation on the flooding was clear: because the system interfacial tension (0.078×10^{-2} mN/m) used for flooding experiment was higher than that of system viscosity of 20 mPa s, it was apparent that the flooding process was in the situation of high capillary number "class-II" displacement state in a large range and for a long time, and the wettability was converted during the flooding process.

It is necessary to find the remedy method after understanding the problem: adjust the combination flooding residual oil saturation to the number similar to that of Model No. 1 and then adjust the wettability conversion parameter T_o for a re-fitting. The fitting figures or digits, stratified residual oil saturation, and related capillary number in the AS-14 tag line (in boldface) in Tables 3.7 and 3.8 were all taken with reference to the fitting result of the core AS-15 experiment. The T_o values of three intervals were adjusted to 2, 4, and 7 (the fitting value T_o of other experiments all took 0.12). In the column "amendment of Model 3" in Table 3.9, the calculation results of corresponding schemes are shown. The recovery efficiency of water flooding scheme was relatively 1% lower than that of the Model No. 0 calculation scheme. The corresponding maximum deviation of the combination flooding was 1.39% for the system scheme No. 1, and the maximum deviation of EOR was 0.42%. The maximum deviation of the recovery efficiency was 2.78% for the system scheme No. 2, and the maximum deviation of EOR was 1.77%. Because this model has the T_o value determined by the fitting calculation, the calculation result of system scheme No. 3 has more reference value, and the relatively small enhanced recovery value may be more accurate.

3.4.3 Determination of digital geology model

Whereas there were three candidate models with more ideal fitting effect, the in-depth comparison was made to the three models in this chapter. The experiment scheme of Xingerxi Block Area and the corresponding water displacement scheme were carried out in Model No. 0, Model No. 1, Model No. 2 (amended), and Model No. 3 (amended). The results are listed in Table 3.10.

As can be seen in Table 3.10, the calculation result of the schemes was very close to that of the scheme on Model No. 0, and the deviation of their water displacement recovery efficiencies was 0.46%, 0.18%, and 0.27%, respectively. The maximum relative error for deviation of Model No. 1 was less than 1% and the minimum relative error for deviation of Model No. 2 (amended) was less than 0.5%. The deviation of compound displacement recovery efficiency was 0.82%, 0.35%, and 1.84%, respectively. The maximum relative error for deviation of Model No. 3 (amended) was less than 3% and the minimum relative error for deviation of Model No. 2 (amended) was less than 0.5%. The deviation of the EOR was 0.36%, 0.17%, and 1.57%, respectively, and the maximum relative error for deviation of Model No. 3 (amended) was 6% or so and the minimum relative error for deviation of Model No. 2 (amended) was less than 0.7%. After the analysis of residual oil condition of the low-permeability layer, the deviation of average residual oil in the whole reservoir was 1.52%, 0.3%,

TABLE 3.10 Flooding effect of Xingerxi experiment scheme on four models.

Scheme	Displacement mode	Implementation time (d)	Maximum underground viscosity (mPa s)	Maximum grid capillary number	Maximum underground pressure specific value	Recovery efficiency (%)	EOR (%)	Stratified residual oil (%)		
								Upper layer	*Middle layer*	*Lower layer*
Experiment	Water flooding					47.20		46.02	41.33	35.67
	Combination flooding	1510	31.04	0.2738	1	72.45	25.25	31.92	17.96	14.42
Model 1	Water flooding					46.74		43.60	39.64	35.00
	Combination flooding	1490	31.01	0.2738	0.9823	71.63	24.89	30.40	17.95	14.64
Model 2 (amended)	Water flooding					46.93		44.95	40.03	34.51
	Combination flooding	1470	29.32	0.2737	0.8425	70.61	23.68	32.67	18.55	14.92
Model 3 (amended)	Water flooding					47.02		44.50	40.18	34.54
	Combination flooding	1480	29.33	0.2738	0.8977	72.10	25.08	31.62	17.32	13.86

and 0.75%, respectively. The deviation of Model No. 1 was the largest and the deviation of Model No. 2 (amended) was the smallest.

The preceding data shows that many indicators for the result of Model No. 2 (amended) were relatively excellent and this model should be selected. For caution's sake, the in-depth analysis was made to the results of Model No. 2 (amended). At the termination of the compound displacement scheme, the residual oil data and distribution situation at the low-permeability position of the oil reservoir are important for testing calculation accuracy. It is shown in Table 3.11 that the residual oil data at the low-permeability position of the oil reservoir when the experiment scheme of Xingerxi carried out on Model

TABLE 3.11 Distribution of low-permeability residual oil at termination of Xingerxi experiment scheme on 3 models (%).

Xingerxi		1	2	3	4	5	6	7	8	9
Experiment Model 0	1	15.7	15.8	16.0	16.2	16.8	19.2	34.2	39.5	50.6
	2	15.8	15.9	16.0	16.3	16.8	19.4	32.8	39.8	50.3
	3	16.0	16.0	16.2	16.5	17.0	20.5	32.6	40.4	49.7
	4	16.2	16.3	16.5	16.8	17.5	23.2	33.6	41.4	49.5
	5	16.8	16.8	17.0	17.5	19.8	25.9	35.9	42.4	49.2
	6	19.1	19.3	20.4	23.1	26.0	32.0	40.2	43.7	48.9
	7	34.0	32.6	32.4	33.4	35.9	40.9	44.0	43.6	48.6
	8	39.5	39.8	40.3	41.2	42.4	43.7	43.6	44.0	47.3
	9	50.6	50.3	49.6	49.4	49.2	48.9	48.7	47.4	47.8
Model 2(Amended)	1	15.8	15.9	16.1	16.3	17.0	20.6	35.3	40.3	50.3
	2	15.9	16.0	16.1	16.4	17.1	21.3	34.7	40.7	50.0
	3	16.1	16.1	16.3	16.6	17.4	23.1	35.9	41.5	49.2
	4	16.3	16.4	16.6	17.0	18.1	25.4	39.1	41.7	48.7
	5	17.0	17.1	17.3	18.1	22.3	31.6	41.4	41.9	48.5
	6	20.5	21.2	22.9	25.5	31.6	39.2	43.2	41.9	48.1
	7	35.3	34.5	35.8	39.0	41.8	41.5	42.0	42.1	46.7
	8	40.3	40.7	41.5	41.8	41.9	42.2	42.3	42.3	45.0
	9	50.3	50.0	49.2	48.7	48.5	47.7	46.8	45.1	45.5
Model 3 (Amended)	1	15.2	15.3	15.5	15.8	16.5	21.2	34.2	38.6	49.2
	2	15.3	15.4	15.5	15.8	16.6	21.8	33.1	38.8	48.9
	3	15.5	15.5	15.7	16.1	16.9	22.7	33.1	39.5	48.4
	4	15.8	15.8	16.1	16.5	18.0	24.9	35.0	40.7	47.8
	5	16.5	16.6	16.9	18.0	22.1	29.0	37.8	41.3	47.5
	6	21.1	21.8	22.4	24.8	28.9	32.9	40.5	41.8	46.9
	7	34.1	32.9	33.0	34.6	37.7	40.5	42.5	41.7	46.6
	8	38.5	38.7	39.4	40.6	41.3	41.8	41.7	41.9	45.0
	9	49.1	48.8	48.3	47.8	47.5	47.0	46.7	45.1	45.4

No. 2 was terminated, and displayed the distribution condition of residual oil in low-permeability reservoir. It is particularly noted that a closed coring well "Xing 2-2—Test Well No. 1"[6] was drilled at 50m behind the oil well on the mainstream line when the field experiment of Xingerxi was terminated. It was found that there was still high saturated crude oil in the core. The location of the inspection well is indicated in the table, where the oil wall peak point has rich compound system displacement. The calculation result coincided with the inspection well. Table 3.11 shows that the calculation results of Model No. 2 (amended) in a comparison form and the position of inspection well was also the oil wall peak point with rich compound system displacement and the oil saturation of 42.5% was relatively high. Upon analyzing the water well pattern, the residual oil of the two schemes was found to be 15.7% and 15.2%, respectively. The residual oil of the two schemes for the oil well pattern was found to be 47.8% and 45.4%, respectively. The residual oil of the two schemes for edge grid at two sides was found to be 50.6% and 49.2%, respectively. The analysis shows that the calculation result of Model No. 2 (amended) was very close to that of Model No. 0.

The comparison curve of Xingerxi Block Area is plotted in Fig. 3.2, which displays the test center well water-cut change curve, relative water displacement EOR change curve, and fitting displacement experiment water-cut change curve. The comparison curve acquired from the displacement scheme of Xingerxi Block Area calculated based on Model No. 2 (amended) is also plotted in Fig. 3.2.

Through more rigorous assessment and screening, Model No. 2 (amended) was determined as the digital geology model.

3.4.4 Summary of technical essentials of establishing digital geology models

The preceding analyses and studies are used to summarize the technical essentials of establishing digital geology model methods. The result of water displacement experiment models shall try the best to comply with the actual situation of the oil reservoir, namely, the water displacement recovery target must be close to the setting targets, and the maximum deviation must be less than 1%. It is recommended to use a two-stage slug for the compound displacement scheme, with the former stage compound system slug to be 0.3 PV and the subsequent polymer slug volume to be 0.5 PV. The slugs of two stages should take the same system viscosity; 20 mPa s is recommended. The appropriate compound system interfacial tension shall be selected according to the research needs. If the displacement experiment goal belongs to a high-viscous and ultra-low interfacial tension system, it is recommended that the system interfacial tension take a value within $2.5–7.5 \times 10^{-3}$ mN/m to complete the displacement experiments. The core wettability conversion parameter *To* during the fitting calculation shall take the default value of 0.12. The stratified reservoir residual oil saturation of compound displacement shall be adjusted and determined. If the displacement experiment goal belongs to a high-viscous and extra-ultra-low interfacial tension system, it is necessary to firstly complete the fitting calculation of ultra-low interfacial tension system experiment to determine the stratified residual oil saturation of compound displacement. Then, take the appropriate extra-ultra-low interfacial tension system to complete the displacement experiment, in which the fitting calculation shall use the residual oil saturation value determined by the previous fitting calculation and adjust and calculate the amended and stratified core wettability conversion parameters *To*. The assessment and screening of the models is relatively easy. In general, the model with ideal displacement experiment results is selected as the experiment model.

3.5 Application of digital geology model of displacement experiment

In the oilfields without implementation of compound displacement experiments, the digital geology model established by the fitting displacement experiments was used for the digital displacement experiments and to design and optimize the displacement experiment schemes so as to speed up the studies and applications of compound displacement technology.

3.5.1 Discussion on producer-injector spacing and scheme slug structure

Large producer-injector spacing is usually used in the preliminary tests of oilfield compound displacement. Producer-injector spacing of 250 m used in the experiments at Daqing Oilfield and other oilfields is not uncommon, and the compound system slug with big volumes is usually used in the experiment schemes to "prevent the failure of compound system surfactant and ensure the good displacing effect." The subsequent small polymer slug was used to prevent water-phase fingering. In order to clarify the problem, a group of schemes was calculated.

The producer-injector spacing was 250 m and total volume of two-stage slugs was 0.5 PV. The slug length of compound system was changed to calculate the different displacement scheme. The annual injection volume was 0.15 PV and the surfactant concentration was 0.3%. In order to highlight the impact of the change of compound system slugs, the same polymer concentration of 2300 mg/L and a system viscosity underground retention rate of 30% were taken. The injection pressure limit of the schemes "borrowed" the pressure limit value P1 acquired from the fitting calculation of Xingerxi Block Area. The main technical data and calculation results of the schemes are listed in Table 3.12.

The system stability and scheme safe implementation issues shall be firstly researched. Table 3.13 shows the low-permeability reservoir residual oil

TABLE 3.12 Flooding program technical data when well spacing is 250 m.

Scheme	Slug volume (PV)		Implementation time (d)	Maximum underground viscosity (mPa s)	Maximum average pressure specific value	Recovery percentage (%)	EOR (%)	Stratified residual oil (%)			Surfactant consumption (t)	Equivalent of total chemical consumption P (t)	Oil increase for ton equivalent polymer (t/t)
	Combinational system	*Polymer*						*Upper layer*	*Middle layer*	*Lower layer*			
1.1	0.20	0.30	2320	34.76	0.976	72.61	25.61	30.49	17.41	13.72	85.44	306.1	83.24
1.2	0.25	0.25	2190	34.95	0.961	72.89	25.89	30.47	16.84	13.70	106.8	338.1	76.18
1.3	0.30	0.20	2276	34.83	0.952	73.02	26.02	30.34	16.67	13.70	128.1	370.1	69.94
1.4	0.35	0.15	2266	34.05	0.929	73.05	26.05	30.34	16.62	13.70	149.5	402.1	64.45
1.5	0.40	0.10	2276	35.01	0.927	73.03	26.03	30.34	16.65	13.71	170.7	434.2	59.64
1.6	0.45	0.05	2281	34.50	0.934	73.01	26.01	30.34	16.68	13.71	192.1	466.2	55.50

TABLE 3.13 Low-permeability residual oil distribution and surfactant concentration distribution when the scheme approached to termination moment (%).

Xingerxi (Review)		1	2	3	4	5	6	7	8	9
Scheme 1.3 S_o (%)	1	15.1	15.2	15.4	15.6	16.1	19.6	32.8	37.6	48.1
	2	15.2	15.3	15.4	15.6	16.2	19.3	30.3	37.7	48.1
	3	15.4	15.4	15.5	15.8	16.4	20.6	31.0	38.2	47.7
	4	15.6	15.6	15.8	16.1	17.0	22.2	32.4	39.2	46.7
	5	16.1	16.2	16.4	17.0	19.9	26.1	35.7	40.0	46.6
	6	19.6	19.7	20.3	22.7	25.6	31.3	39.8	40.3	46.6
	7	32.7	30.6	30.7	32.0	35.4	39.9	40.0	40.6	46.1
	8	37.5	37.7	38.1	39.0	40.0	40.2	40.6	40.8	44.6
	9	48.0	47.9	47.4	46.6	46.5	46.6	46.1	44.8	45.4
Scheme 1.3 C_s (%)	1	0.001	0.004	0.018	0.058	0.104	0.112	0.073	0.028	0.007
	2	0.004	0.01	0.028	0.073	0.11	0.112	0.073	0.028	0.009
	3	0.018	0.028	0.061	0.099	0.115	0.103	0.064	0.027	0.01
	4	0.058	0.072	0.099	0.116	0.114	0.089	0.056	0.027	0.012
	5	0.103	0.109	0.115	0.114	0.098	0.072	0.047	0.026	0.013
	6	0.113	0.111	0.104	0.089	0.073	0.055	0.037	0.025	0.014
	7	0.074	0.073	0.065	0.056	0.047	0.037	0.028	0.025	0.017
	8	0.028	0.028	0.028	0.027	0.026	0.025	0.025	0.024	0.016
	9	0.007	0.009	0.011	0.012	0.013	0.014	0.017	0.015	0.013
Scheme 1.6 S_o (%)	1	15.1	15.2	15.4	15.6	16.2	19.4	33.4	37.9	48.4
	2	15.2	15.3	15.4	15.7	16.4	20.5	31.5	37.9	48.3
	3	15.4	15.4	15.6	15.9	16.6	21.2	31.3	38.3	47.8
	4	15.6	15.7	15.9	16.2	17.1	23.8	32.4	39.0	47
	5	16.2	16.3	16.5	17.2	20.1	26.5	33.6	39.5	46.6
	6	19.4	20.4	21.1	23.7	26.6	30.4	35.1	39.8	46.6
	7	33.3	31.3	31.1	32.2	33.4	35.1	38.6	40.4	46.4
	8	37.8	37.8	38.2	38.9	39.4	39.7	40.4	41.1	45.3
	9	48.5	48.3	47.6	46.9	46.5	46.5	46.3	45.6	46.4
Scheme 1.6 C_s (%)	1	0.002	0.008	0.045	0.145	0.191	0.162	0.084	0.029	0.007
	2	0.008	0.025	0.093	0.179	0.191	0.153	0.083	0.03	0.009
	3	0.044	0.093	0.174	0.200	0.181	0.133	0.076	0.03	0.011
	4	0.145	0.179	0.200	0.190	0.156	0.110	0.068	0.031	0.012
	5	0.190	0.190	0.18	0.156	0.123	0.09	0.061	0.033	0.014
	6	0.163	0.153	0.133	0.111	0.091	0.072	0.055	0.033	0.016
	7	0.085	0.084	0.077	0.069	0.062	0.055	0.044	0.030	0.022
	8	0.029	0.03	0.031	0.032	0.033	0.033	0.029	0.026	0.022
	9	0.007	0.009	0.011	0.013	0.015	0.017	0.022	0.016	0.014

distribution and surfactant concentration distribution when some schemes implemented for 2200 days. The surfactant concentration distribution data for Schemes 1.3 and 1.6 in column 2 and column 4 in Table 3.13 were researched, and the red digits indicate the grid in which the concentration was higher than 0.1%, where the last residence part of the main-body slug of the compound system is. The slug volume of compound system for Scheme 1.6 increased from 0.30 PV to 0.45 PV compared to Scheme 1.3, and in the oil reservoir, the main-body part of the latter slug only moved forward a grid, totaling three grids for outer expansion. Then, the main-body part expanded outward six grids. No significant forward movement appeared for the compound system slug in Scheme 1.6 relative to Scheme 1.4, but in a slow and gradual manner. Thus, if surfactant failure appeared during the displacement process in the scheme of slug volume 0.3 PV, the effective displacement scope would be less than the effective displacement scope of current scheme. The surfactant failure also appeared when the scheme of slug volume 0.45 PV was adopted, and the effective displacement scope of the scheme would also not be significantly increased compared to the scheme of slug volume 0.3 PV. Therefore, the increase of compound system slug volume could not be relied on to ensure surfactant stability in the oil reservoir, and it could only require the displacement process to have an appropriate time limit and must ensure the stability of the surfactant in the whole implementation time. Chapter 2 proposed that the time limit for surfactant stability in the implementation period was 1570 days. From the table, we can see that the shortest implementation time was 2190 days, which dramatically exceeded the safety limit. In order to ensure safe implementation of the displacement experiment, the compound displacement must be implemented with relatively small well spacing.

Given that there was no problem of surfactant stability, the displacement effect problem was subject to further research. Table 3.12 shows that the recovery percentage of the scheme increased by 0.41% when the slug length of the compound system increased gradually from 0.2 to 0.3 PV. The recovery percentage of the scheme, however, decreased by 0.01% when the slug length increased gradually from 0.3 PV to 0.45 PV. Table 3.13 shows the low-permeability residual oil distribution conditions of Scheme 1.3 and Scheme 1.6 for 2200 days of injecting the compound systems. Compared to the system surfactant concentration distribution at the same moment, the last residence part of the compound system slug main body and its rear are residual oil saturation area with a low displacement of compound system. Its front is the oil wall part resided by crude oil displaced from the rear area. The analysis indicated that it did not improve the displacement effect even if the slug of the compound system was too large. A high-efficiency displacement effect can be achieved by adopting the appropriate volume of the compound system slug and the subsequent polymer slug of big enough volume to inhibit the fingering and protect the slug of the compound system.

3.5.2 Determination of key parameters and optimization and design of field experiment scheme

In order to ensure the reliable implementation of the schemes, achieve a higher recovery efficiency, and ensure the termination of the experiments in a short time, the displacement scheme was designed to use the five-spot pattern "four injection wells and nine production wells" with producer-injector spacing of 125 m.

The experiment adopted the scheme recommended in Chapter 2. The slug volume of the compound system was 0.3 PV, surfactant concentration was 0.3%, slug volume of subsequent polymer was 0.6 PV, and injection velocity was 0.4 PV/y. In the scheme, the compound system and the subsequent polymer slug have the same polymer concentration. For the determination of this concentration, it is required to know two parameters relating to target oil reservoir of the experiment: injection pressure limit and displacement system underground viscosity retention rate. The two data are still lacking for the given scheme.

Chapter 2 showed that the maximum reservoir average pressure P during the displacement experiment was closely related to the compound system, the polymer concentration in polymer slug, and the volume of polymer slug, and it played a decisive role in the safe implementation of the schemes and the displacement effect. The maximum reservoir average pressure P of the displacement scheme and the injection pressure limit could only be determined via the fitting of displacement experiments. There was no suitable experiment for the oil reservoir of the given scheme. The injection pressure limit used for the scheme design could not be directly "borrowed" and the "trial" data could be determined only through experiments and with reference to the experiment data of the oil reservoir of similar conditions.

The specific value or ratio between the underground viscosity of displacement system polymer solution and the laboratory preparation viscosity was called "underground viscosity retention rate," and the increase of it is bound to improve the economic benefit of the experiments. If the system viscosity retention rate under test conditions can be confirmed before developing the displacement scheme, the viscosity-concentration curve used in the scheme calculation can adopt the system viscosity retention rate of the same test conditions. In such a case, it can enable the concentration of system polymer solution injected on the site and the viscosity of the solution in field oil reservoir to be consistent with the calculated result, and the field polymer consumption to be consistent with the calculated consumption. It is necessary to acquire the relatively accurate underground viscosity retention rate prior to the scheme design. If it is determined that the retention rate is low, the viscosity retention rate of the system can be improved by taking the measures of selecting the shear-resistant polymer, enlarging perforating density of well low oil reservoirs, perforation aperture, depth, and so on to strive for a viscosity retention rate of greater than 30%.

Because of the "uncertainty" of the two parameters, it is necessary to research the impact of the changes of the two parameters on the displacement effect. The calculation results of displacement schemes are listed in Table 3.14.

The recommended displacement Scheme 1.1 was calculated on the digital geology model established at Xingerxi Block Area. It should be noted that the injection velocity of the scheme is 0.40PV/y, the viscosity retention rate is 30%, and the maximum reservoir average pressure limit was P1 during the displacement process. The displacement Scheme 1.2 was carried out on the target reservoir digital geology model established by the fitting of displacement experiment, and various technical indicators were the same as Scheme 1.1. From the table, it can be learned that the maximum underground system viscosity was relatively low during the implementation process, so the maximum average pressure of the oil reservoir was relatively low. The recovery percentage of the compound displacement decreased by 1.68% and the EOR percentage decreased by 0.89% (Note: the deviation of two models of water displacement recovery percentages was 0.79%). The calculation results show that the two models are very close, and the target reservoir digital geology model established by the fitting of displacement experiment can serve as the design basis of the scheme. Given that the maximum reservoir average pressure limit was P1, Scheme 1.3 was calculated by adjusting the system viscosity. From the table, we learn that the recovery percentage decreased by 1.40% and the EOR percentage decreased by 0.61%, which are closer to Scheme 1.1. This is the design scheme that "accurately offers" the maximum average pressure limit of target reservoir and system viscosity retention rate and the ideal optimization displacement scheme of the target reservoir.

It is not easy to determine the maximum pressure limit of target oil reservoir, and it is very difficult to offer the accurate system viscosity retention rate. The impact of the deviation of the two data on the displacement effect was calculated and researched. Based on Scheme 1.3, the pressure limit value was decreased by 20% to calculate Scheme 1.4. The maximum underground system viscosity relatively decreased by 6.2mPas, declined range by 18.8%; the EOR percentage decreased by 1.42%, declined range by 4.92%. Based on Scheme 1.3, the retention rate of underground system viscosity was decreased to 25% to calculate Scheme 1.5. The maximum underground system viscosity of Scheme 1.5 decreased by 5.54mPas, declined range by 16.8%, and the EOR percentage decreased by 1.37%, declined range by 4.74%. Based on Scheme 1.3, the pressure limit value was decreased by 20%, the retention rate of underground system viscosity was decreased to 25% to calculate Scheme 1.6. The maximum underground system viscosity of Scheme 1.6 decreased by 10.62mPas, declined range by 32.3%, and the EOR percentage decreased by 2.72%, declined range by 9.42%. It is clear that the values of the two indicators significantly affect the displacement effect. The pressure limit deviation and the reduction deviation of underground system viscosity retention rate were then reduced to calculate the three schemes. The injection pressure limit of Scheme

TABLE 3.14 Result of compound flooding experiment in small well spacing.

Scheme	Implementation time (d)	Maximum underground viscosity (mPa s)	Maximum underground pressure specific value	Recovery percentage (%)	EOR (%)	Stratified residual oil (%)			Surfactant consumption (t)	Equivalent of total chemical consumption P (t)	Oil increase for ton equivalent polymer (t/t)
						Upper layer	*Middle layer*	*Lower layer*			
1.1	1170	32.75	0.997	76.74	29.50	24.39	15.92	13.88	32.00	123.2	59.50
1.2	1135	31.89	0.935	75.06	28.61	26.86	15.87	13.40	32.00	123.2	57.70
1.3	1125	32.92	0.998	75.34	28.89	26.51	15.46	13.34	32.00	125.7	57.10
1.4	1140	26.72	0.802	73.92	27.47	28.20	16.84	13.65	32.00	115.8	58.94
1.5	1100	27.38	0.250	73.97	27.52	28.13	16.81	13.64	32.00	125.7	54.40
1.6	1127	22.30	0.802	72.62	26.17	29.55	18.14	13.92	32.00	115.8	56.15
1.7	1140	30.38	0.900	74.83	28.38	27.14	16.04	13.48	32.00	121.0	58.28
1.8	1140	30.18	0.935	74.78	28.33	27.19	16.08	13.49	32.00	125.7	56.00
1.9	1140	27.83	0.900	74.27	27.82	27.84	16.47	13.60	32.00	121.0	57.17

1.7 decreased by 10% compared to that of Scheme 1.3 and the EOR percentage of Scheme 1.7 decreased by 0.51%, declined range by 1.76%. The retention rate of underground system viscosity of Scheme 1.8 decreased to 27.5% and the EOR percentage decreased by 0.56%, declined range by 1.94%. The pressure limit value of Scheme 1.9 decreased by 10% compared to that of Scheme 1.3 and the retention rate of underground system viscosity decreased to 27.5% and the EOR percentage decreased by 1.01%, declined range by 3.70%. The calculation results show that the displacement scheme with a relatively high accuracy could be obtained when the deviation value was in an appropriate range.

As for the oilfields suitable for the application of compound displacement technology and without the implementation of any displacement experiment, the relatively high-quality displacement scheme can be completely designed through the fitting of displacement experiments to establish the digital geology model and through in-depth studies of digital displacement experiments.

3.6 Conclusions

(1) The water displacement and compound displacement experiments have been completed successively by carefully making the 3D core models in compound with the oil reservoir condition. The "equivalent fitting" of the displacement experiment is also carried out under the condition of oil reservoir models to establish the digital reservoir geology model platform, on which the digital displacement experiments are carried out.

(2) It is necessary to establish the oil reservoir model correctly, describing the heterogeneity of the oil reservoir with a simple structure, and then conducting the 3D displacement experiment based on this model. The model precision requires that the deviation between the water displacement experiment recovery efficiency and the recovery efficiency confirmed by oil reservoir shall be lower than 1%; the experiment foundation of establishing the digital geology model of oil reservoir is to use the appropriate system and the reasonable slug structure for compound displacement experiments on the basis of high-precision water displacement experiment.

(3) The geology model of simplified structure shall be established under the oil reservoir condition to carry out the equivalent fitting of laboratory displacement experiments. The key factor for the equivalent fitting is that there is an equal average seepage velocity on the cross-section of two models. The fitting calculation must use the compound displacement software with relatively perfect displacement mechanism. The reservoir geologic data shall be determined through the fitting of water displacement process and the related information data of oil reservoir and compound displacement shall be determined through the fitting of compound displacement process to establish the digital geology model;

(4) The digital geology model shall be used for in-depth studies of digital displacement experiments to increase the laboratory experimental research level and application scope. As for the oilfield suitable for the application of compound displacement technology and without the implementation of any displacement experiment, the digital displacement experiments shall be applied to research and design the relatively high-quality displacement scheme on the basis of carefully researching and determining the key technical data under the reservoir conditions of displacement experiments.

Symbol descriptions

V	seepage velocity of displacement phase, m/s;
μ_w	viscosity of displacing phase, mPa s;
σ_{ow}	interfacial tension between displacing phase and displaced phase, mN/m;
N_c	capillary number value, dimensionless;
N_{ct1}	limit capillary number when conversion of driving condition occurs during combination flooding process, dimensionless;
N_{ct2}	limit capillary number when the residual oil value in combination flooding does not decrease any longer under "Type I" driving condition, dimensionless;
S_{or}^H	lowest residual oil saturation in combination flooding process under "Type I" driving condition, i.e., residual oil saturation corresponding to capillary number lies between limit capillary number N_{ct2} and N_{ct1};
"type I" driving status	displacement under the condition that the capillary number is less than or equal to the limit capillary number N_{ct1};
"type II" driving status	displacement under the condition that the capillary number is higher than the limit capillary number N_{ct1};
T_1, T_2	displacing condition conversion parameter, which can be obtained by experiment or experimental fit;
V_k	reservoir heterogeneity coefficient of variation.

References

1. Lianqing Q. *Numerical Simulation Research of Polymer Displacement Engineering*. Beijing: Petroluem Industry Press; 1998.
2. Fenglan W, Xiaolin W, Guangyu C, et al. Technical progress of alkaline-surfactant-polymer displacement (ASP) in Daqing oilfield. *Petrol Geol Oilfield Development Daqing*. 2009;28 (5):154–162.
3. Demin W, Jiecheng C, Junzheng W, et al. Summary of ASP pilots in Daqing oil field. In: *SPE Asia Pacific Improved Oil Recovery Conference, 25–26 October, Kuala Lumpur, Malaysia* Society of Petroleum Engineers; 1999. https://doi.org/10.2118/57288-MS.

4. Qi LQ, Liu ZZ, Yang CZ, et al. Supplement and optimization of classical capillary number experimental curve for enhanced oil recovery by compound displacement. *Sci China Tech Sci.* 2014;57:2190–2203.
5. Lianqing Q, Hongqing Z, Yanping S, et al. Recommended low-concentration surfactant system for ASP displacement. *Petrol Geol Oilfield Development Daqing.* 2010;29(3):143–149.
6. Shikui L, Yan Z, Yongsheng Z, et al. Evaluation of pilot results of alkali-surfactant-polymer displacement in Daqing oilfield. *Acta Petrol Sin.* 2005;26(3):56–63.

Chapter 4

Construction of microscopic oil/water distribution model of experimental core

Lianqing Qi[a,b], Yanjun Yin[b], Yu Wang[c], DaoShan Li[d], Weihong Qiao[e], Fang Li[b], Junhui Zhang[b], and Yali Wu[b]

[a]*CNPC Daqing Oilfield Exploration and Development Research Institute, Daqing, China,* [b]*CNOOC Energy Technology & Services-Drilling & Production Technology Services Co., Tianjin, China,* [c]*CNPC Xinjiang Oilfield Experiment and Detection Institute, Xinjiang, China,* [d]*CNPC Dagang Oilfield Production Technology Institute, Tianjin, China,* [e]*Chemical Engineering College of Dalian University of Technology, Liaoning, China*

Chapter outline

4.1 Introduction

In the mid-20th century, in order to investigate and describe "the relations between the required hydrodynamic force for the trapped residual oil to flow and capillary retention force" during flooding, American scholars including Moore and Slobod,[1] Taber,[2] and Foster[3] proposed a concept of ratio of

Development and Application of Classical Capillary Number Curve Theory
https://doi.org/10.1016/B978-0-12-821225-7.00004-7

hydrodynamic force to capillary force, also known as capillary number, which is defined via the following formula:

$$N_c = \frac{V \cdot \mu_w}{\sigma_{ow}} \tag{4.1}$$

where N_c stands for capillary number (dimensionless), V for displacing phase seepage velocity (m/s), μ_w for displacing phase viscosity (mPa s), and σ_{ow} for interfacial tension (mN/m) between displacing phase and displaced phase. Relation curves between capillary number and residual oil saturation are then derived from further experiments, briefly known as "capillary number curves." Scholars obtained curves in different forms through researches conducted from different perspectives. The experimental curve in Fig. 4.1 is from Moore and Slobod.

This important research result was published more than half a century ago and marked the start of theoretical research and application of chemical flooding technology. The authors of this volume studied the research findings of the American scholars before entering the chemical flooding research field themselves. Through years of concentrated research, the authors published a paper entitled "Supplementation and Perfection of Classical Capillary Number Experiment Curve for Combination Flooding Enhancing Oil Recovery."[4] The experiments we performed led to the creation of capillary number experiment curve QL as shown in Fig. 4.2. If QL is divided into two parts by limiting capillary number N_{ct2}, the left part looks like the classic capillary number experiment curve as shown in Fig. 4.1; both have approximate corresponding key capillary number. The right part is exactly the supplementation and perfection given to the curve shown in Fig. 4.1.

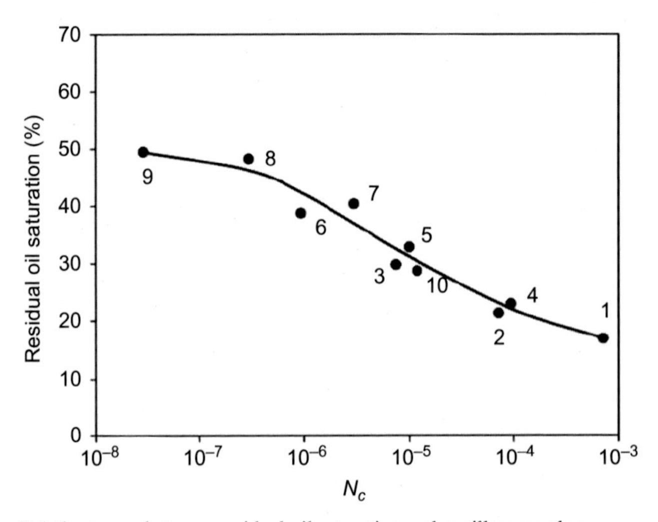

FIG. 4.1 Relation curve between residual oil saturation and capillary number.

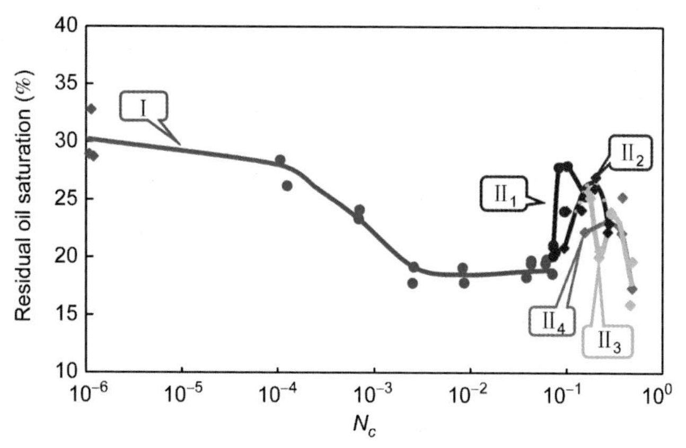

FIG. 4.2 Relation experiment curve *QL* between capillary number and residual oil saturation.

Ref. 4 was published in the journal *Science China Technological Sciences* and attracted concerns and serious attention from peers. However, due to the wide popularity of the classic capillary number experiment curve, the new curve is not easily accepted. The problem focused on "how to interpret the complex forms of the capillary number experiment curve *QL*. "In their paper, the authors introduced the mercury intrusion data of experimental core after the experiment as well as the analysis of injected fluid volume and produced fluid volume of the flooding experiment. They concluded that in a flooding process with higher than limiting capillary number N_{ct1}, where the displacement has reached the oil-water coexistance pore throat zone of the tiny pore canals, some water-wet pore throats show wettability reversal under the higher capillary number displacement condition; the bound water in pore throats is activated and driven away, which produces formation water. These pore throats are transformed to be oil-wet and then trap and turn flowing crude oil to irreducible oil, so that the produced oil decreases and the residual oil saturation increases. It's not easy to make such a conclusion. After the capillary number experiment curve *QL* was obtained, it was immediately reported to the authors' former employer (Exploration and Development Institute of Daqing Oilfield) for verification. According to the understanding at that time, an optimal scheme—one core sample and two displacements—was proposed. First, the displacement is carried out in the case that capillary number is greater than limiting capillary number N_{ct1} to obtain a high residual oil saturation. Then, the displacement is conducted in the case that capillary number is between limiting capillary numbers N_{ct2} and N_{ct1} to further drive out some crude oil and obtain limiting residual oil saturation S_{or}^{H}. The experiment conducted by the Research Institute of Exploration and Development of Daqing Oilfield proves that the displacement done in the case that capillary number is greater than limiting capillary number N_{ct1} leads to high residual oil saturation, but secondary displacement fails to drive out any crude oil. Later, one of the authors personally visited the test team of No.1 Oil

Production Plant of Daqing Oilfield and conducted the experiment and obtained the same results. The authors carefully drew lessons from the failed experiment and gained in-depth knowledge.

After the technical paper[4] was published, the authors continued their in-depth researches. The new research findings are introduced in this chapter.

4.2 The saturating processes of oil and water in the oil displacement experiment core

Capillary number experiment curve is obtained from core flooding experiment, and the microstructure and oil-water distribution in cores determines experimental results. In order to correctly understand oil-water distribution in the core before the flooding experiment, we first introduce the water and oil saturating process in the core.

In order to make the core displacement experiment more similar to the oil displacement process in the actual reservoir, the artificial cores should be similar to the reservoir core conditions and the oil and water distribution in the reservoir should be similar to that in the core samples. To accomplish this, the oil displacement experiment has strict requirements for the saturating oil process in cores. The procedures are roughly as follows:

(1) Use the vacuum pumping method to saturate the cores with water; (2) use the oil displacing water method to saturate oil; and (3) after saturating water and oil, age the postsaturating core for no less than 12 h, in which the oil-water distribution is more uniform and can be used for flooding experiments.

The vacuum pumping method ensures that the core pores are full of connate water. During displacing water by injecting oil, the crude oil enters the pores with a big pore radius first and then moves gradually to the pores with a relatively small pore radius, forming an oil storage network space V_o (Figs. 4.3 and 4.4). With the continuous injection of crude oil, the oil storage space

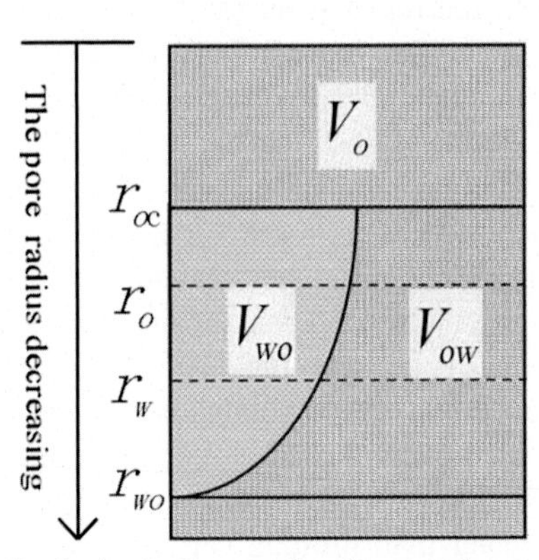

FIG. 4.3 The oil-water distribution in the core after the saturating process (oil-wet).

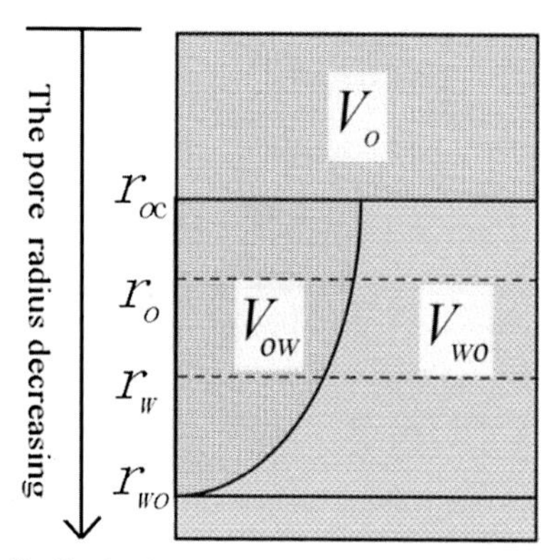

FIG. 4.4 The oil-water distribution in the core after the saturating process (water-wet).

gradually expands to the space with pore radius less than r_{oc}. From the capillary number experiment curve, it can be seen that there is a critical point N_{cc} before and after which the curve shows turning changes. Such changes are related to the core's internal structure, while pore radius r_{oc} is the separation point of the changes. In the space with pore radius greater than r_{oc}, the effects of aperture changing on the capillary resistance gradient of oil flow is relatively small. While the oil flows into the space with pore radius smaller than r_{oc}, the effects of pore diameter changing on capillary resistance gradient of oil flow increase, and the forcefully injected oil can only enter the pores with a relatively smaller radius than r_{oc} with a relatively small resistance so that its flow can be easier. These pores are originally "pure-water" space where flowing oil comes in and gradually expands its "territory" to separate the original "pure-water" space into two kinds of space: "pure-oil" subspace V_{ow}, formed by flowing oil, and "pure-water" subspace V_{wo}. They distribute in a scattered way, mutually separated and surrounded, coexisting in the pores with a pore radius between r_{oc} and r_{wo}. Such pores are referred to as "oil-water coexistence" space. In such an "oil-water coexistence" space, there is bigger interfacial area with a certain amount of surface energy. This is a result of external force during the oil's access. Different wettability of cores determines oil-water distribution in pores with a pore radius less than r_{wo}. For water-wet cores, oil cannot easily enter pores due to action of capillary force so that such pores remain "pure-water" space. For oil-wet cores, oil enters pores during oil saturating due to action of capillary force so that such pores become "pure-oil" space. Fig. 4.3 shows the oil and water distribution in the oil-wet core after the saturating process. Fig. 4.4 shows the oil and water distribution in the water-wet core after the saturating process.

4.3 Microscopic initial oil-water distribution in flooding experiment core

Core samples should undergo 12-h aging treatment after saturating. According to the second law of thermodynamics, substances always tend to reduce any free energy. In a core where oil is saturated, within the "oil-water coexistence" space, a large area of oil-water interface stores high enough surface energy. An aging process that observes the second law of thermodynamics tries to reduce the excessive free surface energy and oil-water surface area. The reduced oil-water surface energy is transformed into capillary force that serves as the driving force to drive oil drops remaining at some places; then the viscous force is transformed into resistance. When driving force is greater than resistance, oil drops begin to flow again and thus help water flow. Oil-water flow also tries to minimize free surface energy so as to follow the second law of thermodynamics. Activated oil drops get close to each other to form a unit before flowing towards and becoming a part of "pure-oil" space V_o. The expanded "pure-oil" space has a minimal pore radius r_o. Scattered flowing water gathers together to eventually form a large-volume "pure-water" space located near the "pure-oil" space. Such a "pure-water" space has a pore radius varying from r_o to r_w. This "pure-water" space is also known as the "water membrane." For oil drops in pores with a pore radius less than r_w, capillary force derived from interface energy cannot meet the activation requirements, so such oil drops remain unchanged, and the pores with a pore radius between r_w and r_{wo} remain "oil-water coexistence" subspace. Oil-water distribution in pores with a smaller pore radius remains unchanged during core aging.

In an aged core, oil-water distribution reaches a new balance. The pores with a pore radius greater than r_o constitute "pure-oil" space V_o, those with a pore radius between r_o and r_w constitute "pure-water" space V_w, and those with a pore radius between r_w and r_{wo} constitute "oil-water coexistence" space. Those with a smaller pore radius constitute either "pure-water" or "pure-oil" space. In the following research, this part of space is not marked separately. If it is "pure-oil" space, it belongs to "pure-oil" subspace V_{ow}. If it is "pure-water" space, it belongs to "pure-water" subspace V_{wo}. Fig. 4.5 shows the oil-water distribution of oil-wet core after saturating oil and aging treatment. Fig. 4.6 shows the oil-water distribution of water-wet core after saturating oil and aging treatment.

4.4 The displacement in different oil storage pore space corresponds to different sections of the capillary number curve

Under the oil and water distribution conditions shown in Fig. 4.5, the driving conditions of different sections of the experimental capillary number curve is analyzed and studied.

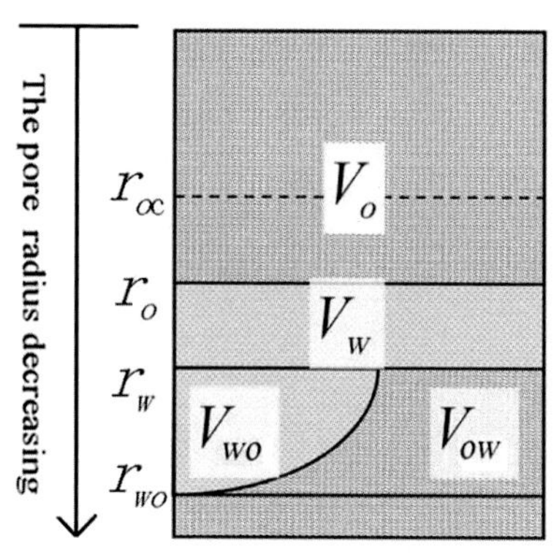

FIG. 4.5 Oil-water distribution of oil-wet core after aging treatment.

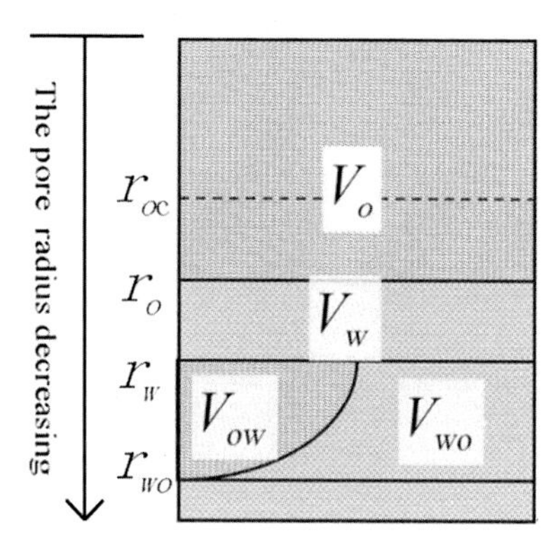

FIG. 4.6 Oil-water distribution of water-wet core after aging treatment.

4.4.1 Displacement under the condition that the capillary number is lower than the limit capillary N_{ct1}

4.4.1.1 Displacement under the condition that capillary number is lower than the limit capillary number N_{ct2}

"Pure-oil" space V_o has a relatively big pore radius. It should also be noted that there is a hidden threshold in space—pore radius r_{oc}. As shown in Fig. 4.7, in the case of water flooding under the condition that capillary number N_c is less than limiting capillary number N_{cc}, the oil in the pores with a pore radius greater than r_{oc} in pore space V_o is driven away during flooding, and we can get a definite residual oil saturation S_{or}. As capillary number increases, the radii of the pores decrease in which crude oil is activated and driven away during flooding. The oil in the pores with a pore radius greater than r_{oc} is driven away and the residual oil saturation S_{or} is the limiting value S_{or}^L when the capillary number reaches the limiting capillary number N_{cc}. As the capillary number further increases during

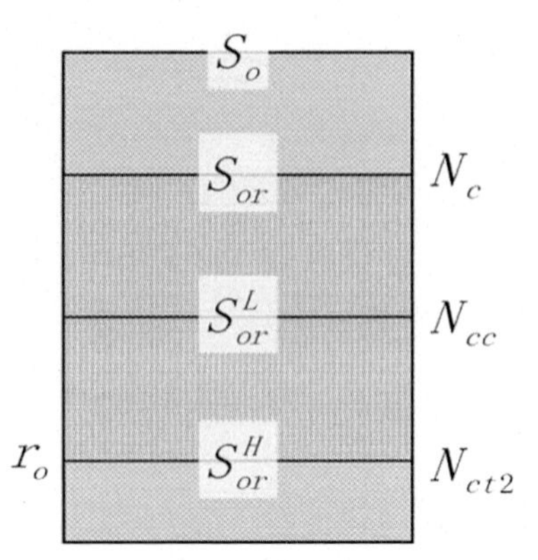

FIG. 4.7 Oil-water distribution while capillary number is less than limiting capillary number N_{cc}.

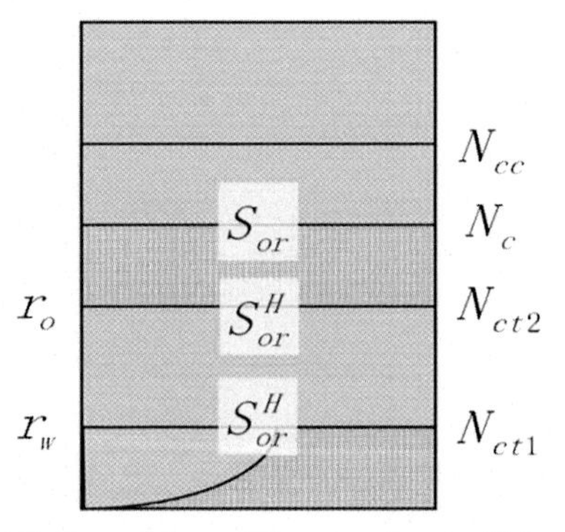

FIG. 4.8 Oil-water distribution while capillary number is higher than limiting capillary number N_{cc}.

flooding, as shown in Fig. 4.8, the pore radius of the space where oil can be driven away is further decreased to a level less than r_{oc}. From the core oil and water saturation process, it can be seen that the capillary force on both sides of the pore radius r_{oc} has a turning point change. At this time, it is necessary to significantly increase the capillary number change gradient to drive out the crude oil in the pores. The displacement experiment officially entered the compound flooding, and the capillary number increased and the residual oil saturation decreased. When the capillary number reaches the limiting value N_{ct2}, the oil in the space with a pore radius greater than r_o is completely driven out while the residual oil saturation S_{or} gradually decreases to the limiting value S_{or}^H.

The experimental data related to curve I in Table 4 of Ref. 4 are listed in the following Table 4.1.

The data in Table 4.1 indicates that in the two water flooding experiments on cores 3-9 and 3-16, the fluid injection velocity is 1.0^{-1} mL/min and the capillary number is around 2.91×10^{-8}, corresponding to residual oil saturation of

TABLE 4.1 Basic data of flooding experiment for capillary number curve (1).

Core no.	Injection velocity (mL min^{-1})	System viscosity (mPa s)	Interfacial tension (mN m^{-1})	Seepage velocity (10^{-5} m s^{-1})	Water-flooding recovery factor (%)	Ultimate recovery factor (%)	Capillary number	Residual oil saturation (%)	Curve mark
3-9	1.00	0.60	2.25×10^{1}	2.919	60.5	60.50	1.09×10^{-6}	28.6	I
3-16	1.00	0.60	2.25×10^{1}	3.180	63.1	63.10	1.19×10^{-6}	28.7	
3-94	0.70	5.30	1.50×10^{0}	3.010	44.4	64.32	1.06×10^{-4}	28.4	
3-8	0.60	6.40	1.20×10^{0}	2.350	41.6	67.31	1.25×10^{-4}	26.2	
3-10	0.60	3.56	1.11×10^{-1}	2.150	37.2	70.29	6.89×10^{-4}	23.3	
3-73	0.60	3.56	1.11×10^{-1}	2.200	37.2	68.50	7.04×10^{-4}	24.1	
3-84	0.60	20.40	1.70×10^{-1}	2.150	45.6	75.00	2.58×10^{-3}	19.2	
3-89	0.60	20.40	1.70×10^{-1}	2.340	46.1	77.61	2.81×10^{-3}	19.9	
3-51	0.80	23.60	8.80×10^{-2}	2.800	0[a]	76.52	7.51×10^{-3}	18.3	
3-63	0.80	23.60	6.50×10^{-2}	2.920	0[a]	76.54	7.83×10^{-3}	18.7	
3-35	0.80	17.90	5.83×10^{-2}	2.700	46.3	74.78	8.29×10^{-3}	19.1	
3-61	0.70	24.00	6.50×10^{-2}	2.330	46.3	75.40	8.59×10^{-3}	17.8	
3-29	0.60	11.90	6.40×10^{-3}	2.050	56.7	76.41	3.81×10^{-2}	18.3	
3-70	1.00	11.50	9.80×10^{-3}	3.650	51.1	75.91	4.28×10^{-2}	19.3	
3-83	1.00	11.50	9.80×10^{-3}	3.640	50.8	75.70	4.27×10^{-2}	19.4	
3-30	0.70	24.10	1.00×10^{-3}	2.500	48.1	75.00	6.03×10^{-2}	19.5	
3-27	0.55	26.40	7.20×10^{-3}	1.940	47.9	75.81	7.12×10^{-2}	18.6	

[a] *The data is obtained from experiments of relative permeability curve test, with no water flooding.*

around 28.6%. In the subsequent experiments, it's hard to increase capillary number by increasing velocity, so a method combining velocity, system interfacial tension, and system viscosity adjusting is used. In the experiment on core 3-94, the capillary number is 1.06×10^{-4} and the corresponding residual oil saturation is 28.4%. In the experiment on core 3-8, the capillary number is 1.25×10^{-4}, the residual oil saturation is 26.2%, the limiting capillary number N_{cc} is approximately between 1.06×10^{-4} and 1.25×10^{-4}, and the residual oil saturation S_{or}^{L} is approximately 27%. As capillary number increases, residual oil saturation rapidly declines. When the capillary number is around 7.04×10^{-4}, corresponding residual oil saturation is about 24%. When the capillary number is 2.58×10^{-3}, residual oil saturation declines to 19.2%. When capillary number is 7.51×10^{-3}, residual oil saturation declines to 18.3%. As the capillary number continues to increase, the residual oil does not reduce any more, and almost all the crude oil in space V_o is produced. As defined in Ref. 4, when residual oil saturation is relatively lowest, the capillary number is called the limiting capillary number N_{ct2}; the experimental data shows that N_{ct2} varies from 2.58×10^{-3} to 7.51×10^{-3}.

According to the classic capillary number curve obtained by American scholars from flooding experiments, as capillary number increases, residual oil saturation steadily decreases. When capillary number reaches a critical value N_{cc}, the residual oil saturation decreases in a relatively rapid way. When capillary number reaches the final value N_{ct}, the residual oil saturation decreases to the limiting value S_{or}^{H}. Depending on experimental conditions, the obtained critical capillary number N_{cc} is roughly between 10^{-5} and 10^{-4}, and the final capillary number N_{ct} is roughly between 10^{-3} and 10^{-2}. When the capillary number experiment curve QL is compared to the American scholars' curve, both have the same form and change trend as well as approximate critical capillary number N_{cc} and limiting capillary number N_{ct}. With respect to their research findings, we use the names given by these scholars to corresponding parameters.

4.4.1.2 Displacement under the condition that capillary number is between limiting capillary numbers N_{ct2} and N_{ct1}

As shown in Fig. 4.9, capillary number in the flooding experiment gradually increases from the limiting capillary number N_{ct2} to open up certain pores with a pore radius less than r_o. In that case, the oil drops originally stored or remained in these pores can flow. As the capillary number gradually increases, the radius of opened-up pores gradually declines. When the capillary number increases up to the limiting capillary number N_{ct1}, the space V_w with a pore radius greater than r_w is opened up on the whole. Flowing oil remaining in the opened-up space during flooding is impossible. In the subsequent different experiments in which the capillary number increases, the residual oil saturation does not decline during the corresponding flooding, indicating that increased capillary number fails to activate originally "irreducible oil," the residual oil saturation

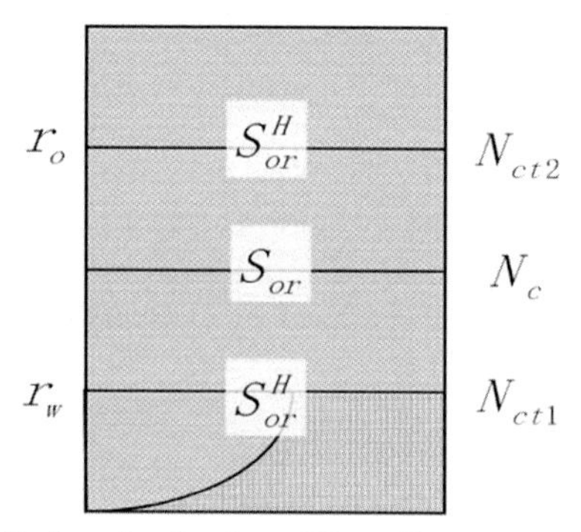

FIG. 4.9 Oil-water distribution after flooding while capillary number is between N_{ct2} and N_{ct1}.

S_{or} stays at S^H_{or}, and the experiment curve shown in the capillary number is flat. Such pore space can only be "pure-water" space.

According to Table 4.1, in the experiment on core 3-84, the capillary number is 2.58×10^{-3}, being the minimum, while the corresponding residual oil saturation is 19.2%. In the experiment on core 3-27, the capillary number is 7.12×10^{-2}, being the maximum, while the corresponding residual oil saturation is 18.6%. Among eleven experiments, the maximum residual oil saturation is 19.7%, the minimum is 17.8%, and the average is 18.76%.

Table 4.2 shows the experimental results of oil displacement with higher capillary number. For core 3-44, the capillary number and the residual oil saturation are 7.20×10^{-2} and 20.1%, respectively; both are relatively minimal. Comparing the data of the core 3-27 experiment with the core 3-44 experiment shows that the capillary number increased slightly from 7.12×10^{-2} to 7.20×10^{-2}, but the residual oil saturation increased significantly from 18.6% to 20.1%. This confirms that the driving status had changed in the oil displacement process of core 3-44. Based on these results, the experimental limiting capillary number N_{ct1} is determined to be 7.12×10^{-2}.

The initial oil-water distribution in the experimental core having a "pure-water" space rationally explains why a flat section that residual oil saturation S_{or} is constant S^H_{or} exists in capillary number experiment curve *QL*. Conversely, a flat section that residual oil saturation S_{or} is constant S^H_{or} existing in capillary number experiment curve *QL* proves that initial oil-water distribution in the experimental core necessarily has a "pure-water" space.

4.4.2 Displacement under the condition that the capillary number is greater than the limiting capillary number N_{ct1}

4.4.2.1 Original oil-water distribution in "oil-water coexistence" space and distribution changes after flooding

When flooding is carried out under the condition that capillary number is greater than limiting capillary number N_{ct1}, oil and water in the "oil-water

TABLE 4.2 Basic data of flooding experiment for capillary number curve (2).

Core no.	Injection velocity (mL min^{-1})	System viscosity (mPa s)	Interfacial tension (mN m^{-1})	Seepage velocity (10^{-5} m s^{-1})	Capillary number	Residual oil saturation (%)	L (%)	Curve mark
3-44	0.60	24.00	7.20×10^{-3}	2.160	7.20×10^{-2}	20.1	1.34	II_1
3-45	0.60	24.00	7.20×10^{-3}	2.170	7.25×10^{-2}	21.1	2.34	
3-18	0.60	24.30	7.20×10^{-3}	2.270	7.65×10^{-2}	20.5	1.74	
3-14	0.55	26.40	7.20×10^{-3}	2.250	8.23×10^{-2}	27.8	9.04	
3-11	0.80	16.50	5.00×10^{-3}	2.860	9.43×10^{-2}	24.0	5.24	
3-53	0.60	29.20	6.80×10^{-3}	2.327	9.99×10^{-2}	22.3	5.24	
3-15	0.80	16.50	5.00×10^{-3}	3.080	1.02×10^{-1}	27.9	9.14	
3-72	0.90	19.50	6.80×10^{-3}	3.290	9.44×10^{-2}	20.8	2.04	II_2
3-23	0.90	28.60	6.80×10^{-3}	3.391	1.43×10^{-1}	23.6	5.34	
3-85	0.90	38.20	6.80×10^{-3}	3.495	1.96×10^{-1}	25.0	7.14	
3-54	0.90	38.30	6.80×10^{-3}	3.566	2.01×10^{-1}	26.9	8.14	
3-17	0.90	53.50	6.80×10^{-3}	3.513	2.76×10^{-1}	22.9	4.14	
3-62	0.90	53.50	6.80×10^{-3}	3.388	2.67×10^{-1}	22.2	3.34	

3-5	0.60	10.90	1.50×10^{-3}	2.420	1.76×10^{-1}	25.1	6.34	II_3
3-69	0.60	10.90	1.50×10^{-3}	2.350	1.71×10^{-1}	25.7	6.94	
3-81	0.60	15.10	1.50×10^{-3}	2.160	2.17×10^{-1}	20.0	1.24	
3-31	0.60	15.10	1.50×10^{-3}	2.210	2.22×10^{-1}	20.6	1.84	
3-3	0.60	21.90	1.70×10^{-3}	2.260	2.91×10^{-1}	23.7	4.94	
3-48	0.60	21.90	1.70×10^{-3}	2.240	2.89×10^{-1}	23.9	5.14	
3-1	0.60	37.40	1.70×10^{-3}	2.190	4.81×10^{-1}	19.6	0.84	
3-26	0.60	37.40	1.70×10^{-3}	2.080	4.57×10^{-1}	15.9	−2.86	
3-50	0.90	7.20	1.50×10^{-3}	3.220	1.55×10^{-1}	22.2	3.44	II_4
3-21	0.90	11.10	1.50×10^{-3}	3.580	2.65×10^{-1}	23.0	4.24	
3-57	0.90	17.30	1.50×10^{-3}	2.460	2.83×10^{-1}	25.2	6.34	
3-82	0.90	17.30	1.50×10^{-3}	3.230	3.72×10^{-1}	22.1	3.34	
3-68	0.90	24.10	1.50×10^{-3}	2.990	4.80×10^{-1}	17.3	−1.46	

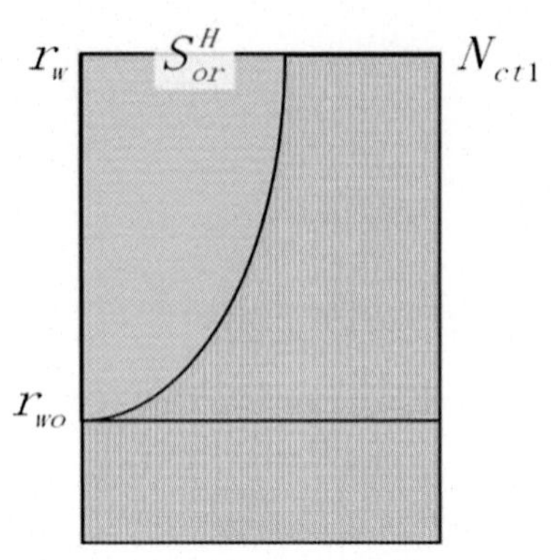

FIG. 4.10 Oil-water distribution in oil-water coexistence space.

coexistence" space begins to be activated. The initial oil-water distribution in "oil-water coexistence" space is separately presented in Fig. 4.10. Hereafter "pure-water" subspace in "oil-water coexistence" space is referred to as "irreducible water" space, and "pure-oil" subspace is referred to as "irreducible oil" space.

As can be seen from Fig. 4.10, in the oil-water coexistence space, within the same aperture range, there is both pure-oil subspace and pure-water subspace. Moreover, the aperture of the oil-water coexistence space is small enough, and capillary force plays a more prominent role. As shown in Fig. 4.11, there is the "bound oil" subspace ① and the "irreducible water" subspace ① when the pore radius is between r_w and r_{wo1}. In the displacement under the condition of the capillary number N_c is slightly higher than the limit capillary number N_{ct1},

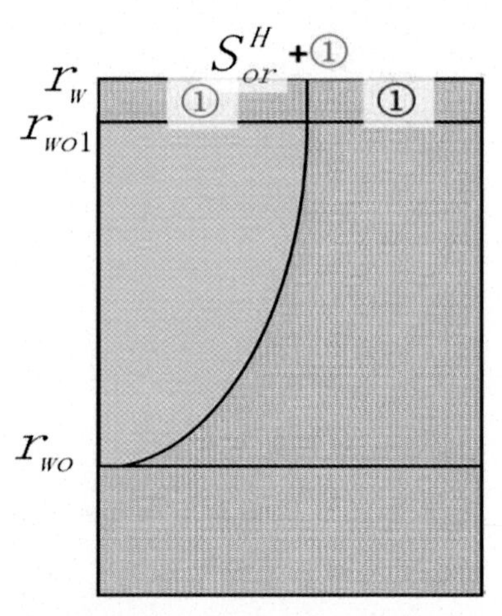

FIG. 4.11 Oil-water distribution after flooding in "oil-water coexistence" space while capillary number is slightly higher than N_{ct1} (1).

the differences of oil-water apparent viscous resistance displayed, the "bound oil" in subspace ① can't be started, and the "irreducible water" in the subspace ① can be started. Therefore the flow oil in the oil-water flow space with the adjacent aperture higher than r_w has the opportunity to enter into the subspace ①. Once the oil droplets enter into the relative fine channel, the flow resistance in the pore increases, the velocity of flowing water is relatively fast, and the oil droplets are easily blocked stuck down. This activation and flowing away requires greater capillary number. However, this space does not meet such conditions, so flowing oil entering space ① is captured to become irreducible oil. L_1 is used to represent volume percentage of captured oil in space ①, namely, the ratio of captured oil volume to core pore volume. Residual oil saturation after flooding is $S_{or}=S_{or}^{H}+L_1$, meaning that the residual saturation gets higher when the oil displacement process terminates.

The capillary number of flooding experiment is further increased. Likewise, as shown in Fig. 4.12, the pore radius of subspace ② activating irreducible oil is between r_w and r_{ow2}; L_2 is used to represent volume percentage of released irreducible oil. While that of subspace ② trapping flowing oil is between r_{ow2} and r_{wo2}. If L_1 is still used to represent volume percentage of trapped flowing oil, the residual oil saturation after the experiment is:

$$S_{or}=S_{or}^{H}+L=S_{or}^{H}+(L_1-L_2) \tag{4.2}$$

The flooding experiments after capillary number is further increased will lead to the aforementioned results repeatedly. Likewise, we can infer what happens to the nth experiment. For the nth experiment, as shown in Fig. 4.13, the pore radius of subspace ⓝ activating irreducible oil is between r_w and r_{own}; L_2 is

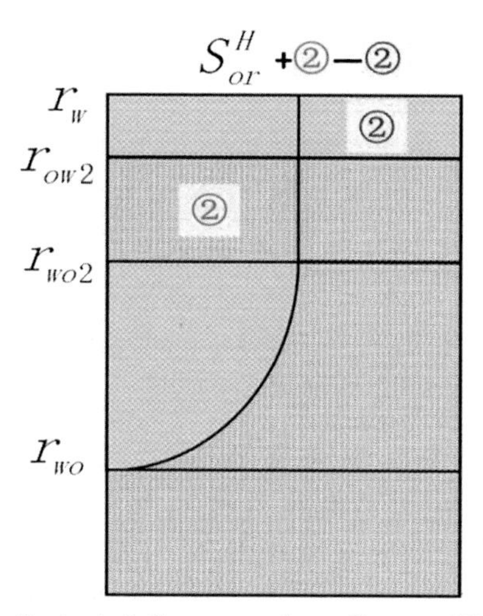

FIG. 4.12 Oil-water distribution in "oil-water coexistence" space while capillary number is relatively higher than N_{ct1} (2).

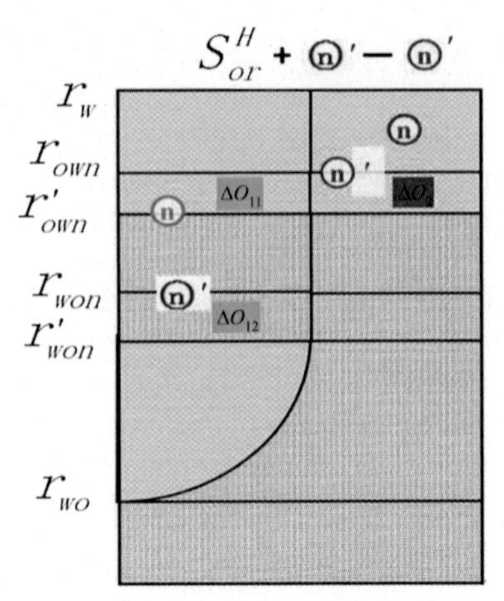

FIG. 4.13 Oil-water distribution in "oil-water coexistence" space while capillary number is greater than N_{ct1} after flooding (3).

used to represent volume percentage of released irreducible oil. While that of subspace ⓝ trapping flowing oil is between r_{own} and r_{won}. If L_1 is still used to represent the volume percentage of trapped flowing oil, the residual oil saturation of the nth flooding experiment is:

$$S_{or} = S^H_{or} + L_n = S^H_{or} + (L_1 - L_2) \tag{4.3}$$

The residual oil saturation S_{or} of the flooding experiment can be determined as can the limiting residual oil saturation S^H_{or}. Thus, the relative increment L of residual oil saturation S_{or} can be determined as well. The increment L comprises two parts: L_1 for volume percentage of trapped flowing oil, and L_2 for volume percentage of released irreducible oil. Although L_1 and L_2 cannot be measured directly, it has been clearly realized that they are relevant key factors leading to increased residual oil saturation.

Consider $n' = n + 1$, during the n'th flooding experiment, that the capillary number relatively increases. Fig. 4.13 shows the correlation quantity mark of n' displacement experiment. The pore radius of the subspace ⓝ' activating irreducible oil is between r_w and $r_{own'}$, the volume percentage of activated oil is represented by L_2', the pore radius of the subspace ⓝ' trapping flowing oil is between $r_{own'}$ and $r_{own'}$, where $r_{own'}$ is less than r_{own} and $r_{won'}$ is less than r_{won}, and L_1' is used to represent volume percentage of trapped oil. The residual oil saturation after the experiment is:

$$S'_{or} = S^H_{or} + L_{n+1} = S^H_{or} + (L'_1 - L'_2) \tag{4.4}$$

The difference between the residual oil saturations from two successive experiments is:

$$\Delta = S'_{or} - S_{or} = (L'_1 - L_2) - (L_1 - L_2) = (L'_1 - L_1) - (L'_2 - L_2) \tag{4.5}$$

As can be seen in Fig. 4.13, oil content $(L_2' - L_2)$ is the oil content in subspace ΔO_2, and oil content $(L_1' - L_1)$ is the difference in oil content between subspace ΔO_{12} and the subspace ΔO_{11}, which can be simplified as:

$$\Delta = \Delta O_{12} - \Delta O_{11} - \Delta O_2 \tag{4.6}$$

From the experimental curve of capillary number, based on limit capillary number N_{ct1}, as capillary number gradually increases, because the residual oil saturation increment Δ between two experiments is greater than 0, the residual oil saturation S_{or} increases gradually, at the appropriate capillary number N_c, $\Delta = 0$, residual oil saturation S_{or} obtain the maximum, thereby Δ turn into less than zero value, as the capillary number increases, the corresponding residual oil saturation is reduced gradually, at a suitable capillary number, residual oil saturation S_{or} is equivalent to limit S_{or}^H again, and as capillary number continue to increase, the residual oil saturation continue to reduce.

Table 4.2 shows the experimental data under the condition that capillary number is greater than limiting capillary number N_{ct1} in experiment curve *QL*; some of experimental data positions are adjusted.

In the first three experiments in curve II_1, the liquid injection velocity and interfacial tension are the same as is the system viscosity. Compared with the core 3-27 experiment, the system interfacial tension is the same, the liquid injection rate is high, and the system viscosity is low. In the process of oil displacement, the capillary numbers are 7.20×10^{-2}, 7.25×10^{-2}, and 7.65×10^{-2} respectively, which are all higher than the capillary number 7.12×10^{-2} of the core 3-27 experiment. At the end of oil displacement process, the residual oil saturation is 20.1%, 21.1%, and 20.5% respectively, which is higher than the residual oil saturation value of 18.6% of the core 3-27 experiment, and higher than the calculated residual oil saturation limit value S_{or}^H 18.76%. It can be considered that in the "bound oil" subspace, no "bound oil" is released and driven away, i.e., $L_2 = 0$, while in the "bound water" subspace ①, the volume percentage of captured oil L_1 is about 1.34%, 2.34%, and 1.74%, and the relative increment L of residual oil saturation S_{or} is equal to L_1. In the oil displacement experiments of core 3-11 and core 3-53, the interfacial tension, injection velocity, and system viscosity are different from each other. The capillary number in the oil displacement process is similar, 9.43×10^{-2} and 9.99×10^{-2}, respectively. The residual oil saturation S_{or} is 24%, and relative residual oil saturation limit value S_{or}^H increment L is 5.24%. Compared with the previous group of experiments, the capillary number in oil displacement process increases, the residual oil saturation value increases, the relative residual oil saturation S_{or}^H limit value increment L increases, and the volume percentage of the captured oil L_1 in the "bound water" subspace ① is higher than L. For the oil displacement experiments of cores 3-14 and 3-15, system interfacial tension, injection speed, the system viscosity are different, the capillary number of displacement process is 8.23×10^{-2} and 1.02×10^{-1}, which is a large difference, but the displacement effect is very close. Residual oil saturation values S_{or} are

27.8% and 27.9%, respectively, and the relative residual oil saturation limit value S_{or}^{H} increment L is 9.04% and 9.14%, respectively. A search for the cause revealed a clear difference in core structure. The oil displacement of core 3-14 and core 3-15 was separated. Seven experiments were sorted as shown in Table 4.2. In the first six experiments, both the capillary number and residual oil saturation increase. The residual oil saturation of the core 3-14 experiment is the maximum value; thereafter the residual oil saturation decreases as the capillary number increases. After the core 3-14 experiment is deducted, another group of six experiments is conducted. With the increase of capillary number, the residual oil saturation gradually increases, and the residual oil saturation of the core 3-15 experiment is the maximum value.

For the experimental curve II_2, the fluid injection velocity is 0.9 mL/min and the system interfacial tension variation is 6.8×10^{-3} mN/m. Six flooding experiments are performed in the case that the system viscosity gradually increases. Analysis of the first five experiments in the table shows that the capillary number gradually increases as does the corresponding residual oil saturation S_{or} initially. The capillary number of the core 3-54 experiment reaches 2.01×10^{-1} and the residual oil saturation 26.9%, which is a relatively maximum value. The experimental capillary number continues to increase later, and the residual oil saturation declines. It should be noted that two parallel experimental data on cores 3-17 and 3-62 are extremely close. The experimental data showed that the residual oil saturation was slightly higher in the case of large capillary number, and "slightly" reversed sequence, check the record to confirm that there are differences in core structure data, which is result in relative large seepage velocity in the core 3-62 experiment, and due to shearing action system viscosity is relatively lower in displacement process, and the capillary number calculation took a relatively high viscosity values. If using the actual viscosity value to calculation, the experiment capillary number may be a little less than 2.67×10^{-1}, so there is no problem order. According to the experimental data, the capillary number of the core 3-54 experiment reaches 2.01×10^{-1}, and the residual oil saturation is 26.9%, which is higher than the limit value of residual oil saturation S_{or}^{H} by 8.14%, seeing the difficulty of oil producing from the oil-water space. The experimental capillary number of core 3-62 reached 2.67×10^{-1}, system viscosity reached 53.50 mPa s, and residual oil saturation was 22.2%, which is higher than the residual oil saturation limit S_{or}^{H} by 3.34%. It is clear that the high interfacial tension of the system is an important factor that leads to higher residual oil saturation. The interfacial tension of the experimental system in this group is 6.80×10^{-3} mN/m, which meets the requirements of industry standards.

4.4.2.2 Further understanding of the experimental curve of displacement capillary Number in "oil-water coexistence" space

In the curve II_4 experiment the fluid injection velocity remains 0.9 mL/min, but the system interfacial tension is adjusted to 1.50×10^{-3} mN/m. Five flooding experiments are performed in the case that the system viscosity changes. According to the order in the table, the system viscosity of the scheme gradually increases as does the capillary number, corresponding to the experimental residual oil saturation changes from gradually increasing to decreasing, and the core 3-57 experiment achieves the maximum value. In the core 3-68 experiment, the viscosity of oil displacement system was 24.1 mPa s, the capillary number in the oil displacement process was 4.80×10^{-1}, and the residual oil saturation value S_{or} was 17.3%, about 1.46% lower than the limit residual oil saturation S_{or}^{H}.

In the curve II_3 experiment the fluid injection velocity is decreased to 0.6 mL/min, and the system interfacial tension of the flooding experiment is basically the same as the previous II_4 experiment. According to the experimental requirements, the displacement fluid was prepared according to the design requirements the afternoon before the oil displacement experiment, and the interfacial tension data of the system was measured. Before the oil displacement experiment, the determination was verified again. The data given in the table are the verification data. It can be seen from the data that the interfacial tension of the high-viscosity system increases relatively, which is a normal phenomenon for an ultra-low interfacial tension system. Analysis of the experimental results shows that the eight core experiments are four groups of parallel experiments. For the 3-5 and 3-69 core displacement experiments, the interfacial tension, viscosity, corresponding displacement speed, and capillary number are similar, 1.76×10^{-1} and 1.71×10^{-1}, respectively. The final residual oil saturation value is similar at 25.1% and 25.7%, respectively. The 3-81 and 3-31 core experiments in the second group were similar to those in the former group. The capillary number in the oil displacement process was 2.17×10^{-1} and 2.22×10^{-1}, respectively. The residual oil saturation value at the termination of the oil displacement process was 20.% and 20.6%, respectively. There are two situations: one is that the capillary number value of first group of experiments is relatively small, obtaining relatively large residual oil saturation value; the other is that no residual oil saturation increases as capillary number increases, which is all normal. The reason is that the capillary number of the first group of experiments is sufficiently large and residual oil saturation has reached a "peak." In the second group of experiments, the capillary number increases relative to the first group of experiments; the residual oil saturation incremental Δ is less than zero and thus the residual oil saturation decreases. When the measured interfacial tension value of the system was 1.70×10^{-3} mN/m, the core displacement experiments of 3-3 and 3-48 cores were completed, and the capillary number in the oil displacement process was 2.22×10^{-1} and 2.20×10^{-1}, respectively. Residual oil saturation was 23.7% and 23.9% when the oil displacement process was terminated. We continued to increase the viscosity of the oil displacement system, and complete the core displacement experiments of 3-1 and 3-26 cores. The capillary number of the oil displacement process is 3.87×10^{-1} and 3.85×10^{-1}, respectively. The residual oil saturation of the oil displacement process is 19.6% and 15.9%, respectively. They have a big deviation, and the latter has dropped below the residual oil saturation value S_{or}^{H}. Two things are worth noting. First, the residual oil saturation of the third group of experiments is higher than that of the second group, and the residual oil saturation of the fourth group of experiments is lower than that of the third group. This is apparently because the residual oil saturation incremental Δ of the third group is greater than 0, relative to the second group, and the residual oil saturation incremental Δ of the fourth group is less than 0, relative to the third group. The curve II_3 experiment shows "twin peaks," and the "twin peaks" form of capillary number curve may be the "standard" form of the experimental curve when the capillary number is higher than the limit capillary number N_{ct1}.

According to the preceding analysis, when the capillary number is higher than the limit value N_{ct1}, the displacement process enters the "oil-water coexistence" space. The experiment made four curves, and each curve corresponds to a certain condition. Residual oil saturation S_{or} is composed of two parts: the volume percentage of oil not started in the "bound oil" subspace, and the volume percentage of capture flow oil in the irreducible water subspace. Residual oil saturation S_{or} can be expressed by a compound function:

$$S_{or} = F(N_c) = F(\mu_w, V/\sigma_{ow}) \tag{4.7}$$

The ratio V/σ_{ow} can be used as an appropriate comparison condition, correspond to different viscosity μ_w of displacement phase, complete a group of oil displacement experiments, and make a corresponding experimental curve.

Through our research we clearly show that residual oil saturation increases or decreases when capillary number is higher than the limit capillary number N_{ct1}. In the authors' original paper,[4] the reason for the increase and decrease of residual oil saturation is explained to be caused by "wettability transformation" "under the condition of high capillary number displacement." Now it seems this is not the case. According to the "wettability transformation," it is improper to name the parameters T_0 as "wettability transformation influence parameters," which should be changed to "driving condition transformation parameters."

4.4.2.3 Carding of correspondence between core microscopic space and capillary number experimental curves

There are three important limiting capillary numbers on the capillary number experiment curve QL. N_{cc} is the limiting capillary number from water flooding to composite flooding, which is the maximum limiting capillary number to drive out oil in the space V_{o1} (pure-oil space V_o with pore radius greater than r_{oc}) by water flooding, and the minimum limiting capillary number to drive oil in the space V_{o2} (pure-oil space V_o with pore radius less than r_{oc}) by composite flooding, which corresponds to the pore radius r_{oc} at the interface between V_{o1} and V_{o2} in the "pure-oil" space V_o. The capillary number N_{ct2} refers to the limiting capillary number when residual oil saturation changes from a declining trend to a flat trend, being the maximum limiting value of capillary number activating "pure-oil" space V_o during flooding and also the minimal limiting value of capillary number activating bound water in "pure-water" space V_w. The limiting capillary number N_{ct2} corresponds to the pore radius boundary value r_o of "pure-oil" space V_o and "pure-water" space V_w. The capillary number N_{ct1} refers to the boundary capillary number when residual oil saturation changes from a flat trend to a complex trend, being the maximum limiting value activating bound water in "pure-water" space V_w during flooding and the minimum limiting value of capillary number activating bound water in "bound-water" subspace V_{wo} and bound oil in "bound-oil" subspace V_{ow} in "oil-water coexistence" space. The limiting capillary number N_{ct1} corresponds to the pore radius boundary value r_w of the two subspaces V_{ow} and V_{wo} of "oil-water coexistence" space and "pure-water" space V_w.

Therefore, it is not difficult to measure core initial "bound water" saturation S_{wr}^L, which is distributed in "pure-water" space V_w and "bound water" subspace V_{wo}, and the initial oil saturation S_o, which is distributed in "pure-oil" subspace V_{o1}, V_{o2} and V_{ow}, with oil content pore volume percentage parameters v_{o1}, v_{o2} and v_{ow} respectively, $S_o = 1 - S_{wr}^L = v_{o1} + v_{o2} + v_{ow}$. The limit capillary number N_{cc} of water drive corresponds to the oil amount in the subspace V_{o1}, and the limit value of residual oil saturation is S_{or}^L, $S_{or}^L = 1 - S_{wr}^L - v_{o2} - v_{ow}$. After that, the capillary number in the compound drive increases to the limit value N_{ct2}, corresponding to the oil amount in the subspace V_{o2}, and the limit value of residual oil saturation is S_{or}^H, $S_{or}^H = 1 - S_{wr}^L - v_{o1} - v_{o2} = v_{ow}$. As the capillary number N_c continues to increase, displacement enters the oil-water coexistence space, and the residual oil saturation value S_{or} presents complex changes.

4.5 Similarities and differences in microcosmic space of reservoirs with different permeability

In literature,[4] the assessment and test of limit capillary number N_{ct2} and N_{ct1} was conducted on the core with different permeability produced by referring to the reservoir conditions of Daqing Oilfield, under oil-water condition of Daqing Oilfield. Here, further discussion is presented based on the micro-spatial oil-water distribution model of the core.

Base on Table 6 in Ref. 4, some relevant data is supplemented to constitute Table 4.3. According to the data shown in Table 4.3, the flooding experiments on three cores give very approximate limiting capillary numbers N_{ct2} and N_{ct1}. Further review of the experimental data shows that the value ranges of the two parameters can be further shrunk for the latter two groups of experiments. It can be therefore concluded that the pay zones with different permeability have nearly the same limiting capillary numbers N_{ct2} and N_{ct1}. Three groups of cores have nearly the same limit capillary number N_{ct2}, which means that their "pure-water" space V_w and "pure-oil" space V_o have nearly the same pore radius and

TABLE 4.3 Key data of flooding experiments on different core groups for capillary number curve.

Core group	Average K (10^{-3} μm^2)		N_{ct1}		N_{ct2}		Finally recovery factor (%)	$S_o - S_{or}^H$ (%)	S_{or}^H (%)
	Gas measured	*Water measured*	*Lower limit*	*Upper limit*	*Lower limit*	*Upper limit*			
1	2839	871	7.12×10^{-2}	7.20×10^{-2}	7.04×10^{-4}	2.58×10^{-3}	75.74	58.29	18.67
2	2375	784	1.11×10^{-1}	1.62×10^{-1}	5.91×10^{-4}	2.09×10^{-3}	65.77	46.58	24.24
3	1274	460	1.51×10^{-1}	2.13×10^{-1}	5.89×10^{-4}	1.34×10^{-3}	59.53	46.47	31.59

structural characteristics at their boundaries. Likewise, three groups of cores have nearly the same limit capillary number N_{ct1}, meaning their "pure-water" space V_w and "oil-water coexistence" space have nearly the same pore radius and structural characteristics at their boundaries. It is not difficult to explain that the same rock sand with the same material and particle size, the same cement, under the same standard requirements and process conditions, and the same size of rock sand with different dosage ratio, under the same oil and water conditions, the capillary number of oil and water with the same aperture should be the same. It is regretful that no test was conducted on limit capillary number N_{cc}, however, it can be concluded that the one and only three groups of core "pure-oil" space V_o "interface" between subspace V_{o1} and V_{o2} has nearly the same pore radius, the nearby similar pore structure characteristics, they will have similar limit capillary number of N_{cc}.

The experimental data of capillary number experimental curve QL are listed in Tables 4 and 5 in literature,[4] and the capillary number of the flooding processes of twenty-two cores is between the limit capillary number N_{ct2} and N_{ct1}. The experimental data of fourteen oil displacements are selected and listed in Table 4.4 in three groups. Data include core volume, pore volume, porosity, gas measured permeability, saturated oil volume, initial oil saturation, water content volume, and residual oil saturation S_{or}^{H} measured by the experiment.

It can be seen from the data in the table that the deviation between "each group of data" in the same column is relatively small. For the convenience of comparison, the mean value is calculated first, and then the corresponding data is calculated under the condition that the same volume of core is 587 cm^3.

In the first column, cores are homogeneous with relatively high permeability, and core average values are as follows: pore volume is 166 mL, porosity = 28.36%, overall saturated oil capacity = 130 mL, oil saturation S_o is 78.31%, space V_o pore volume is 98.0 mL, accounting for 59% of the percentage value v_o of pore volume, and pore volume of V_{ow} is 31.6 mL, accounting for 19% of total pore volume that can be regarded as the residual oil saturation limit S_{or}^{H}. The initial water is 36.4 mL, accounting for 22% of the pore volume. By comparison, this group of cores has high pore volume, high porosity, high saturated oil content, high oil saturation, large volume space V_o, and small volume space V_{ow}. Since the crude oil in space V_o can be extracted when the capillary number is higher than the limit N_{ct2}, and the space V_{ow} crude oil can only be extracted when the capillary number is much higher than the limit N_{ct1} and the displacement fluid has relatively high viscosity, it can be seen that this kind of core composite flooding is less difficult to extract, has high economic and technical benefits, and is more suitable for composite flooding.

In the second column, cores are homogeneous with relatively low permeability, and core average values are as follows: pore volume is 152 mL, porosity is 25.83%, the overall saturated oil capacity is 119 mL, oil saturation is 78.26%, space V_o pore volume is 71.6 mL, v_o value is 47.1%, space V_{ow} pore volume is 47.4, S_{or}^{H} is 31.2%, and the initial water volume is 33 mL, accounting for 21.7%

TABLE 4.4 Capillary number experiment curve *QL* experiment data table (3).

Core number	Core volume (cm^3)	Pore volume (mL)	Porosity (%)	Saturated oil volume (mL)	Permeability (gas measured) (μm^2)	S_o (%)	S_{or}^H (%)	V_o space volume (mL)	V_{ow} space volume (mL)	Initial water content volume (mL)
3-70	591	165	28.22	132	2851	80.00	19.3	100.2	31.8	33
3-63	582	165	28.43	130	2867	78.79	18.4	99.2	30.8	35
3-61	590	164	27.80	128	2843	78.05	19.9	95.4	32.6	36
3-30	590	168	28.48	131	2812	77.98	18.7	99.6	31.4	37
3-27	591	168	28.42	129	2824	76.79	18.6	97.8	31.2	39
3-84	579	167	28.82	128	2843	76.65	19.2	96.0	32.0	39
Mean	**587**	**166**	**28.36**	**130**	**2840**	**78.31**	**19.0**	**98.0**	**31.6**	**36**
Y-25	311	80	25.67	61	1213	76.25	30.5	36.6	24.4	19
Y-8	311	79	25.35	63	1269	79.75	32.7	37.2	25.8	16
Y-5	311	82	26.31	66	1245	80.49	31.5	40.2	25.8	16
Y-3	311	81	25.99	62	1251	76.54	30.1	37.6	24.4	19
Mean	**311**	**81**	**25.83**	**63**	**1245**	**78.26**	**31.2**	**37.9**	**25.1**	**18**
Isopyknic	**587**	**152**	**25.83**	**119**	**1245**	**78.26**	**31.2**	**71.6**	**47.4**	**33**
S-23	577	138	23.90	99	2369	71.74	23.6	66.4	32.6	39
S-29	572	138	23.69	102	2342	73.91	24.6	68.0	34.0	36
S-16	580	140	24.14	100	2355	71.43	24.0	66.4	33.6	40
S-22	575	138	24.01	102	2376	73.91	24.8	67.8	34.2	36
Mean	**576**	**139**	**23.94**	**101**	**2361**	**72.75**	**24.3**	**67.2**	**33.6**	**38**
Isopyknic	**587**	**141**	**23.94**	**102**	**2361**	**72.75**	**24.3**	**67.9**	**34.1**	**39**

of the pore volume. Compared with the first group, the core of this group has low pore volume and corresponding low porosity and low saturated oil content. However, due to the relatively low pore volume, it has high oil saturation, space V_o has relatively small volume, and space V_{ow} has relatively large volume. This type of core composite flooding is difficult and has low economic and technical benefits.

In the third column, cores are heterogeneous with three layers, and core average volumes are as follows: pore volume is 141 mL, porosity is 23.94%, the overall saturated oil capacity is 102 mL, oil saturation S_o is 72.75%, space V_o pore volume is 67.9 mL, v_o value is 48%, space V_{ow} pore volume is 34.1 mL, S_{or}^H value is 24.2%, and the initial water is 39 mL, accounting for 27.7% of the pore volume. Compared with the former two groups, this group has lower core pore volume and corresponding lower porosity, lower saturated oil and corresponding lower oil saturation, space V_o has relatively minimum volume, space V_{ow} volume is relatively higher than that of the first column of the core, and lesser relative to the second column core. The difficulty of this kind of core flooding is greater and has lower economic and technical benefits.

Three groups of cores with different permeability have nearly the same capillary number limit N_{ct2} and N_{ct1} and significantly different limit residual oil saturation. Further study showed that there were significant differences between the "pure-oil" space V_o and "pure-oil" space V_{ow} of the three groups of cores. It was the "pure-oil" space V_{ow} that determined the significant difference in the limit residual oil saturation of the three groups of curves. The volume percentage v_{ow} of oil content in "pure-oil" space V_{ow} is equal to S_{or}^H.

Reservoirs with different permeability are formed under the same diagenetic conditions by using the same materials and sand grains of rocks with different grain sizes, and stores oil and water of the same nature. The preceding research results are applicable to this kind of reservoir and beneficial to the research and application of compound oil displacement technology. The data can be directly applied to the design of a geological model in digital oil displacement test research.

4.6 The displacement of different oil storage pore in water-wet cores corresponds to different section of the capillary number curve

Section 4.2 studies the saturated water-oil process of the core before the oil displacement experiment. For oil-wet cores and water-wet cores, they are the same. However, due to the opposite wettability of rock sand in the process of oil saturation, there is an obvious difference in saturated oil amount. Obviously, the saturated oil content of the oil core is relatively large. Moreover, during the "aging" treatment of the core after saturated oil, the distribution of oil and water in the microscopic space is adjusted to "reduce the interface energy." The hydrophilic and hydrophilic characteristics of rock sand are different, which also leads to the pore radius difference at the interface of microscopic "pure-oil" space and "pure-water" space. Two important differences between micro oil and water distribution of wet core and oil core before oil displacement experiment should be clearly recognized.

In oil displacement, Section 4.3 shows that as the capillary number gradually increases, the displacement process gradually progresses to the pore space with finer aperture. These understandings are also applicable to water-wet cores.

When the capillary number N_c is lower than the limit capillary number N_{ct1}, the displacement process takes place in the "pure-oil" space V_o and "pure-water" space V_w. Their pore radius is large enough, and wettability has relatively little influence on oil driving. The corresponding relationship between displacement in different oil storage spaces and different sections of capillary number curve of water-wet cores is the same as that of oil-wet cores. When the capillary number is higher than the limit N_{ct1}, the displacement process enters the "oil-water coexistence" space. With the decrease of pore radius, the effect of wettability gradually increases. It should be noted that for water-wet core there are two conditions: the pore radius is less than the r_w, coexist with same aperture corresponding to the oil "bound" pore space and the bound water pore space, and the presence of a certain difference in oil and water viscosity, the capillary number of starting the same aperture "bound oil" is higher than that of bound water.

In Fig. 4.14, in the displacement with limit capillary number N_{ct1}, the "bound oil" can't be started in subspace ① with the pore radius in $r_w \sim r_{wo1}$, while in the subspace ① with same aperture, the bound water can flow. The flow oil in the pore space ① is captured to become bound oil, which is in common with the oil-wet cores. The water-wet core orifice has a relatively strong "displacement" ability for oil droplets, which corresponds to a relatively weak

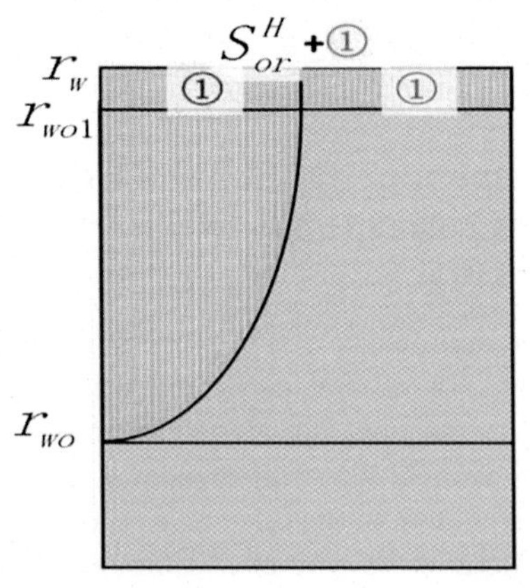

FIG. 4.14 Oil and water distribution diagram at the end of displacement process when the capillary number of "oil-water coexistence" space is N_{ct1}.

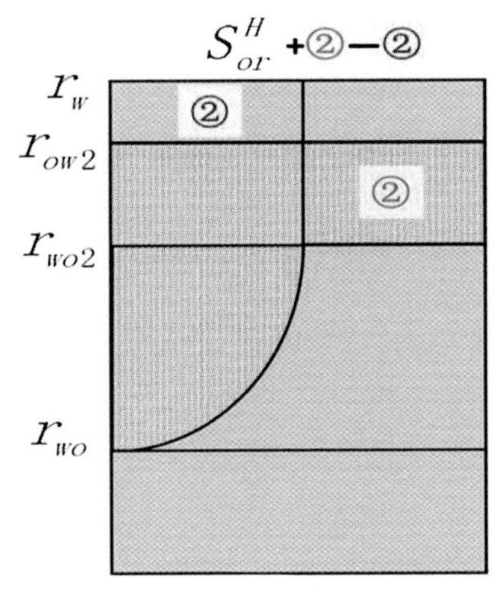

FIG. 4.15 Oil and water distribution diagram at the end of displacement process when the capillary number of "oil-water coexistence" space is higher than N_{ct1}.

"capture" ability for flowing crude oil. It can be assumed that the capillary number has the same increment on the basis of the limit capillary number N_{ct1}, and the capturing flow oil L_1 is relatively small. The lower limit pore radius r_{wo1} of space ① is relatively increased, the residual oil saturation value S_{or} ($=S^H_{or}+L_1$) is relatively smaller. Increase the capillary number in flooding experiment, see Fig. 4.15. It is assumed that the increase of capillary number N_c relative to limiting capillary number N_{ct1} is the same as that in the case of oil wetting. In the case of water wetting, the external action required to start "bound oil" is relatively small, so the lower limit of pore radius r_{ow2} of subspace ② with "bound oil" released is relatively small, and the percentage L_2 of volume of released bound oil is relatively large. At this point, some spaces in the "bound water" subspace with a pore radius higher than r_{ow2} have lost the ability to capture flowing oil, and the lower bound of the pore radius r_{ow2} of subspace ② to capture flowing oil has an increase relative to that of oil wetting, and the volume percentage L_1 of "flowing oil" captured in the subspace ② decreases relatively, and the residual oil saturation ($S_{or}=S^H_{or}+L=S^H_{or}+(L_1-L_2)$) of the experiment decreases relatively.

If the capillary number of oil displacement experiment is increased again, the same situation will reoccur. Therefore, the nth experiment can be summarized, as shown in Fig. 4.16. The pore radius of the subspace ⓝ of the start-up "bound oil" space is between r_w and r_{own}, and the percentage of volume of released "bound oil" is marked as L_2. The pore radius of the subspace ⓝ of the "bound water" space to capture flowing oil is between r_{own} and r_{won}, and the volume percentage of the captured "flowing oil" is marked by L_1. Thus, the residual oil saturation of the nth displacement experiment is obtained:

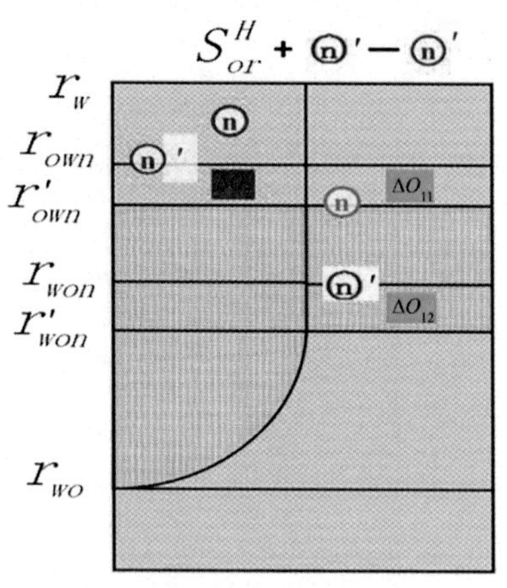

FIG. 4.16 Oil and water distribution diagram at the end of displacement process when the capillary number of "oil-water coexistence" space is higher than N_{ct1} (2).

$$S_{or} = S^{H}_{or} + L_n = S^{H}_{or} + (L_1 - L_2)$$

For the $n' = n + 1$ oil displacement experiment, the capillary number in the oil displacement process increases relatively. The pore radius of the starting "bound oil" subspace ⓝ' is between r_w and $r_{own'}$, and the volume percentage of released oil is L_2'. The pore radius of the subspace ⓝ' of the "bound water" space capturing the flowing oil is between $r_{own'}$ and $r_{won'}$, and the pore radius $r_{own'}$ is less than r_{own}, the pore radius $r_{won'}$ is less than r_{won}, the volume percentage of captured oil is L_1'. Residual oil saturation at the end of the experiment:

$$S'_{or} = S^{H}_{or} + L_{n+1} = S^{H}_{or} + L'_1 - L'_2$$

Residual oil saturation difference for n and n' adjacent experiments:

$$\Delta = \Delta O_{12} - \Delta O_{11} - \Delta O_2$$

If initial incremental Δ is greater than 0, the residual oil saturation S_{or} increases gradually. When $\Delta = 0$, the residual oil saturation S_{or} obtains the maximum, then Δ decreases to a value less than zero, and the corresponding residual oil saturation decreases. At the appropriate capillary value, there is again residual oil saturation S_{or} equivalent to the limit value S^{H}_{or}.

Compared with oil wetness, it can be seen that the variation rules of residual oil saturation of the two types of cores are similar. However, for wet cores, it is relatively easy to release "trapped oil" from the void and relatively difficult to capture "flowing oil." The increase rate of residual oil saturation is relatively slow, the increase of relative residual oil saturation is relatively small, the decline rate of residual oil saturation is relatively fast, and the oil stored in the "oil-water coexistence" space is relatively easy to drive out.

4.7 Conclusions

(1) The model of oil-water distribution characteristics in microscopic space of displacing experiment cores is created according to the basic theory and definition of seepage flow mechanics and by analyzing a large amount of capillary number experimental curve QL data, as well as with the help of strict inferential analysis. The construction of a microcosmic oil-water distribution model of core deepens the understanding of microcosmic space structure and oil-water distribution of the reservoir. A new method for understanding oil and water distribution in microcosmic space of core is established. This achievement will certainly be applied in future oil field exploration and development.

(2) The microscopic distribution of oil and water in a core is orderly arranged. It is sorted in descending order of pore radius, the "pure-oil" space V_o with pore size higher than r_o, "pure-water" space V_w with pore size between r_o and r_w, and "oil-water coexistence" space with pore size between r_w and r_{wo}. If the pore size is smaller than r_{wo}, it is either "pure-oil" space or "pure-water" space, which is determined by core wettability. For the stealth interface with aperture r_{oc}, the "pure-oil" space V_o is divided into two parts. When the aperture is higher than r_{oc}, it is the subspace V_{o1}; when the aperture is lower than r_{oc}, it is the subspace V_{o2}.

(3) There is a corresponding relationship between the pore radius r_{oc} , r_o, and r_w in the core microscopic space oil and water distribution model and the limiting capillary number N_{cc}, N_{ct2}, and N_{ct1} on the capillary number curve. N_{cc} is the limit capillary number to drive out crude oil in of the pure-oil subspace V_{o1}, which corresponds to the pore radius r_{oc} and the limit residual oil saturation S_{or}^L. N_{ct2} is the limit capillary number to drive out crude oil in of the pure-oil subspace V_{o2}, which corresponds to the pore radius r_o and the limit residual oil saturation S_{or}^H. N_{ct1} is the limit capillary number to start oil-water flow in the oil-water coexistence space, corresponding to the pore radius r_w.

(4) When the capillary number N_c is higher than the limit capillary number N_{ct1}, the displacement residual oil saturation presents complex changes. The reasons for this are that in the "oil-water coexistence" space where the pore radius is smaller than r_w, there is a "confined water" space corresponding to the pore radius of "bound oil" space, and the pore radius of the space is small enough. Due to the obvious difference in viscosity between oil and water, the capillary number of starting "bound oil" in the same aperture is higher than that of "bound water." The pore characteristics and oil-water distribution of "oil-water coexistence" space determine the variation law of residual oil saturation.

(5) It can be obtained via oil displacement experiment and reasoning that the cores of reservoirs with different permeability formed under the same diagenetic conditions by rocks and sand grains of the same material with

different particle sizes, and by the same cementing material, store oil and water of the same nature, and form microscopic oil-water space. The corresponding curve of capillary number of chemical drive has nearly the same limit capillary number N_{cc}, N_{ct2}, and N_{ct1}. The difference is that the corresponding space has different pore volumes.

(6) The order of microscopic oil-water subspace of water-wet cores is basically the same as that of oil-wet cores. The corresponding relationship between microscopic oil-water subspace displacement and capillary number experimental curve of water-wet core is the same as that of oil-wet core.

Symbol description

V	displacement speed, m/s
μ_w	displacing phase viscosity, mPa s
σ_{ow}	interfacial tension between displaced phase and displaced phase, mN/m
N_c	capillary number
N_{cc}	limit capillary number when residual oil begins to flow after water flooding and polymer flooding， that is, the limit capillary number of oil flow in pure-oil subspace V_{o2} initiated by compound flooding
N_{ct1}	limit capillary number when the driving condition changes in the process of compound oil driving, that is, the limit capillary number of oil-water flow in oil-water coexistence space initiated by compound flooding
N_{ct2}	the capillary number of the compound flooding displacement out crude oil in "pure-oil" pore space V_o under the condition of "type I" driving status, that is, the limit capillary number when the residual oil saturation no longer decreases under the condition of "type I" driving status
"type I" driving status	displacement under the condition that the capillary number is less than or equal to the limit capillary number N_{ct1}
"type II" driving status	displacement under the condition that the capillary number is higher than the limit capillary number N_{ct1}
S_w	water phase saturation
S_{wr}^L	irreducible water saturation
S_o	oil phase saturation
S_{or}	residual oil saturation corresponding to capillary number N_c
S_{or}^L	under the condition of low capillary number, when the capillary number N_c is less than or equal to N_{cc}, the limit

	value of residual oil saturation is equal to the oil content v_{o1} in the subspace V_{o1} of "pure-oil" pore space V_o
S_{or}^H	in the "I class" driving conditions, minimum residual oil saturation in the displacement process, that is, the residual oil saturation corresponding to the capillary number between the limit capillary number N_{ct2} and N_{ct1}
V_o	the pore space with a micro pore radius greater than r_o is initially "pure-oil" pore space
r_{oc}	pore radius of interface between subspaces V_{o1} and V_{o2} of "pure-oil" pore space V_o
V_{o1}	the "pure-oil" pore space V_o with pore radius greater than r_{oc}
V_{o2}	the "pure-oil" pore space V_o with pore radius less than r_{oc}
r_o	the pore radius of the interface between the microscopic "pure-oil" pore space V_o and the microscopic "pure-water" pore space V_w
V_w	"pure-water" pore space with the pore radius r from r_o to r_w
r_w	pore radius of the interface between "pure-water" pore space V_w and "oil-water coexistence" space
V_{ow}	the "pure-oil" pore space with pore radius r is less than r_w
V_{wo}	the "pure-water" pore space with pore radius r is less than r_w
v_o	pore volume percentage of initial oil content in microscopic pore space V_o (Note: abbreviated "pore volume percentage of oil content" is "oil content."); after reservoir development, v_o volume includes oil volume retained in space V_o and space V_w
v_{o1}	pore volume percentage of initial oil content in microscopic pore space V_{o1}
v_{o2}	pore volume percentage of initial oil content in microscopic pore space V_{o2}
v_{ow}	pore volume percentage of initial oil content in microscopic pore space V_{ow}; after reservoir development, v_{ow} volume includes oil volume retained in space V_{ow} and space V_{wo}

References

1. Moore TF, Slobod RC. The effect of viscosity and capillarity on the displacement of oil by water. *Producers Month.* 1956;20(10):20–30.
2. Taber JJ. Dynamic and static forces required to remove a discontinuous oil phase from porous media containing both oil and water. *Soc Pet Eng J.* 1969;9(1):3–12.

3. Foster WR. A low tension waterflooding process employing a petroleum sulfonate, inorganic salts, and a biopolymer. *SPE*. 1972;3803.
4. Qi LQ, Liu ZZ, Yang CZ, et al. Supplement and optimization of classical capillary number experimental curve for enhanced oil recovery by combination flooding. *Sci China Technol Sci*. 2014;57:2190–2203.

Chapter 5

Further study on the capillary number experimental curve and phase permeability curve

Lianqing Qi[a,b], Qiang Wang[c], Bailing Kong[d], Ronghua Li[e], Liantao Shan[f], Ying Ma[g], Wei He[b], and Duansheng Shi[b]

[a]*CNPC Daqing Oilfield Exploration and Development Research Institute, Daqing, China,* [b]*CNOOC Energy Technology & Services-Drilling & Production Technology Services Co., Tianjin, China,* [c]*CNPC Exploration and Development Institute, Beijing, China,* [d]*SINOPEC Henan Oilfield Exploration and Development Institute, Henan, China,* [e]*CNPC Xinjiang Oilfield Experiment and Detection Institute, Xinjiang, China,* [f]*SINOPEC Shengli Oilfield Exploration and Development Institute, Shandong, China,* [g]*CNPC Huabei Oilfield Company Petroleum Exploration and Production Research Institute, Renqiu, China*

Chapter outline

The capillary number experimental curve *QL* is a supplement and perfection of the classic capillary number experimental curve. By analyzing and studying the oil displacement experiment results of the capillary number experimental curve *QL*, we constructed the "core microscopic oil-water distribution model," which deepened the understanding of the reservoir core microstructure and the

Development and Application of Classical Capillary Number Curve Theory
https://doi.org/10.1016/B978-0-12-821225-7.00005-9

oil-water distribution. Then, through the research on the "core microscopic oil-water distribution model," we deepened the understanding of the capillary number experimental curve, and raised new issues at the same time.

5.1 Two descriptions on the capillary number experimental curve *QL*

The capillary number is defined as follows:

$$N_c = \frac{V \cdot \mu_w}{\sigma_{ow}} \tag{5.1}$$

Wherein, N_c is the capillary number (nondimensional), V is the seepage velocity (m/s), μ_w is the displacing phase viscosity (mPa s), and σ_{ow} is the interfacial tension between the displacing phase and the displaced phase (mN/m).

The seepage velocity of the classic capillary number experimental curve adopts the "displacing velocity," which is defined as follows:

$$V = Q/A \tag{5.2}$$

Q is the volume of injecting liquid of the oil displacement experiment, and A is the core interface pore area.

We started to study alkali-surfactant-polymer (ASP) flooding technology from the numerical simulation calculation research, and further studied the capillary experiment curve. In the calculation study, it is necessary to separately study the water phase capillary number and the oil phase capillary number. Thus, the following formula was used to calculate the velocity V in sorting out the experimental data of the capillary number in the literature[1]:

$$V = \frac{Q}{A \times (1 - S_{or})} \tag{5.3}$$

Wherein, "S_{or}" is the residual oil saturation of the corresponding oil displacement experiment. This formula defines the limit value of the water seepage velocity in the core. By using this velocity, the capillary number value obtained should be the "capillary number of the water seepage" in the displacing process. The experimental data in section 4.1 of the literature[1] were rearranged and listed in Table 5.1. The data sorted out by the formula (5.2) were listed as (type 1) data, and the data sorted out by the formula (5.3) were listed as (type 2) data. Apparently, the "type 1" capillary number is divided by $(1 - S_o)$ to obtain the corresponding "type 2" capillary number. Fig. 5.1 is the curve chart plotted using the "type 1" capillary number, and Fig. 5.2 is the curve chart plotted using the "type 2" capillary number. By contrast, there was no significant difference in the curve shape between the two figures, and only the capillary number of the corresponding experiment was reduced accordingly.

For the purpose of distinguishing, the two curves were named as curve *QL* (I) and curve *QL* (II), and the two curves were used differently. The curve *QL* (I)

TABLE 5.1 Basic data of oil displacement experiment of capillary number curve QL.

Experiment no.	Core no.	Flooding rate (mL min^{-1})	System viscosity (mPa s)	Interfacial tension (mN m^{-1})	Displacement seepage velocity (10^{-5} m s^{-1})	V/σ_{ow} value (10^{-6} mPa s^{-1})	Capillary number (type 1)	Water phase seepage velocity (10^{-5} m s^{-1})	Capillary number (type 2)	Residual oil saturation (%)	Curve label
1	3-9	1.00	0.60	2.25×10^{1}	2.919	1.297	7.78×10^{-7}	4.080	1.09×10^{-6}	28.6	I
2	3-16	1.00	0.60	2.25×10^{1}	3.180	1.413	8.48×10^{-7}	4.460	1.19×10^{-6}	28.7	
3	3-94	0.70	5.30	1.50×10^{0}	2.155	14.31	7.59×10^{-5}	3.010	1.06×10^{-4}	28.4	
4	3-8	0.60	6.40	1.20×10^{0}	1.734	14.45	9.25×10^{-5}	2.350	1.25×10^{-4}	26.2	
5	3-10	0.60	3.56	1.11×10^{-1}	1.649	148.6	5.29×10^{-4}	2.150	6.89×10^{-4}	23.3	
6	3-73	0.60	3.56	1.11×10^{-1}	1.670	150.5	5.36×10^{-4}	2.200	7.04×10^{-4}	24.1	
7	3-84	0.60	20.40	1.70×10^{-1}	1.737	102.2	2.08×10^{-3}	2.150	2.58×10^{-3}	19.2	
8	3-89	0.60	20.40	1.70×10^{-1}	1.874	110.2	2.25×10^{-3}	2.340	2.81×10^{-3}	19.9	
9	3-51	0.80	23.60	8.80×10^{-2}	2.288	260.0	6.14×10^{-3}	2.800	7.51×10^{-3}	18.3	
10	3-63	0.80	23.60	8.80×10^{-2}	2.374	269.8	6.37×10^{-3}	2.920	7.83×10^{-3}	18.7	
11	3-35	0.80	17.90	5.83×10^{-2}	2.184	374.6	6.71×10^{-3}	2.700	8.29×10^{-3}	19.1	
12	3-61	0.70	24.00	6.50×10^{-2}	1.915	294.6	7.07×10^{-3}	2.330	8.59×10^{-3}	17.8	
13	3-29	0.60	11.90	6.40×10^{-3}	1.675	2617	3.11×10^{-2}	2.050	3.81×10^{-2}	18.3	
14	3-70	1.00	11.50	9.80×10^{-3}	2.946	3006	3.46×10^{-2}	3.650	4.28×10^{-2}	19.3	
15	3-83	1.00	11.50	9.80×10^{-3}	2.934	2994	3.44×10^{-2}	3.640	4.27×10^{-2}	19.4	
16	3-30	0.70	24.10	1.00×10^{-2}	2.013	2013	4.85×10^{-2}	2.500	6.03×10^{-2}	19.5	
17	3-27	0.55	26.40	7.20×10^{-3}	1.579	2193	5.79×10^{-2}	1.940	7.12×10^{-2}	18.6	

Continued

TABLE 5.1 Basic data of oil displacement experiment of capillary number curve QL—cont'd

Experiment no.	Core no.	Flooding rate (mL min^{-1})	System viscosity (mPa s)	Interfacial tension (mN m^{-1})	Displacement seepage velocity (10^{-5} m s^{-1})	V/σ_{ow} value (10^{-6} mPa s^{-1})	Capillary number (type 1)	Water phase seepage velocity (10^{-5} m s^{-1})	Capillary number (type 2)	Residual oil saturation (%)	Curve label
18	3-44	0.60	24.00	7.20×10^{-3}	1.723	2403	5.77×10^{-2}	2.160	7.20×10^{-2}	20.1	II_1
19	3-45	0.60	24.00	7.20×10^{-3}	1.712	2378	5.71×10^{-2}	2.170	7.25×10^{-2}	21.1	
20	3-18	0.60	24.30	7.20×10^{-3}	1.805	2505	6.09×10^{-2}	2.270	7.65×10^{-2}	20.5	
21	3-53	0.60	29.20	6.80×10^{-3}	1.808	2659	7.76×10^{-2}	2.327	9.99×10^{-2}	22.3	
22	3-11	0.80	16.50	5.00×10^{-3}	2.174	4348	7.17×10^{-2}	2.860	9.43×10^{-2}	24.0	
23	3-14	0.55	26.40	7.20×10^{-3}	1.625	2257	5.96×10^{-2}	2.250	8.23×10^{-2}	27.8	
24	3-15	0.80	16.50	5.00×10^{-3}	2.221	4442	7.33×10^{-2}	3.080	1.02×10^{-1}	27.9	
25	3-72	0.90	19.50	6.80×10^{-3}	2.606	3832	7.47×10^{-2}	3.290	9.44×10^{-2}	20.8	II_2
26	3-23	0.90	28.60	6.80×10^{-3}	2.591	3810	1.09×10^{-1}	3.391	1.43×10^{-1}	23.6	
27	3-85	0.90	38.20	6.80×10^{-3}	2.621	3854	1.47×10^{-1}	3.495	1.96×10^{-1}	25.0	
28	3-54	0.90	38.30	6.80×10^{-3}	2.607	3838	1.47×10^{-1}	3.566	2.01×10^{-1}	26.9	
29	3-17	0.90	53.50	6.80×10^{-3}	2.709	3984	2.13×10^{-1}	3.513	2.76×10^{-1}	22.9	
30	3-62	0.90	53.50	6.80×10^{-3}	2.634	4982	2.07×10^{-1}	3.388	2.67×10^{-1}	22.2	

31	3-5	0.60	10.90	1.50×10^{-3}	1.813	12086	1.32×10^{-1}	2.420	1.76×10^{-1}	25.1	II_3
32	3-69	0.60	10.90	1.50×10^{-3}	1.746	11640	1.27×10^{-1}	2.350	1.71×10^{-1}	25.7	
33	3-81	0.60	15.10	1.50×10^{-3}	1.728	11520	1.74×10^{-1}	2.160	2.17×10^{-1}	20.0	
34	3-31	0.60	15.10	1.50×10^{-3}	1.755	11700	1.77×10^{-1}	2.210	2.22×10^{-1}	20.6	
35	3-3	0.60	21.90	1.70×10^{-3}	1.724	10141	2.22×10^{-1}	2.260	2.91×10^{-1}	23.7	
36	3-48	0.60	21.90	1.70×10^{-3}	1.705	10029	2.20×10^{-1}	2.240	2.89×10^{-1}	23.9	
37	3-1	0.60	37.40	1.70×10^{-3}	1.761	10359	3.87×10^{-1}	2.190	4.81×10^{-1}	19.6	
38	3-26	0.60	37.40	1.70×10^{-3}	1.749	10288	3.85×10^{-1}	2.080	4.57×10^{-1}	15.9	
39	3-50	0.90	7.20	1.50×10^{-3}	2.505	16700	1.20×10^{-1}	3.220	1.55×10^{-1}	22.2	II_4
40	3-21	0.90	11.10	1.50×10^{-3}	2.757	18380	2.04×10^{-1}	3.580	2.65×10^{-1}	23.0	
41	3-82	0.90	17.30	1.50×10^{-3}	2.516	16770	2.90×10^{-1}	3.230	3.72×10^{-1}	22.1	
42	3-57	0.90	17.30	1.50×10^{-3}	1.840	12267	2.12×10^{-1}	2.460	2.83×10^{-1}	25.2	
43	3-68	0.90	24.10	1.50×10^{-3}	2.473	16487	3.97×10^{-1}	2.990	4.80×10^{-1}	17.3	

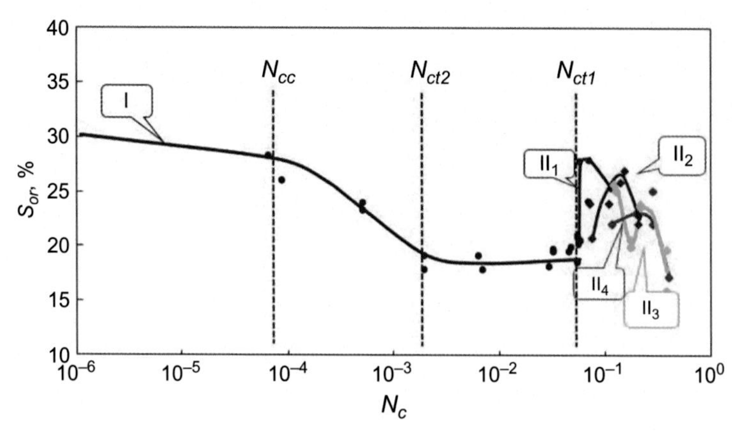

FIG. 5.1 Capillary number experimental curve *QL* (I).

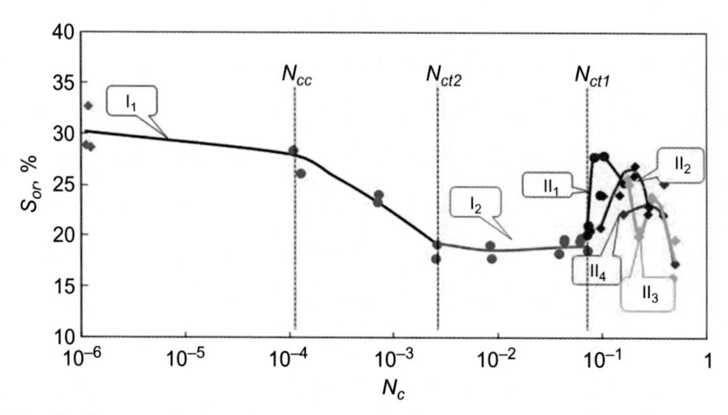

FIG. 5.2 Capillary number experimental curve *QL* (II).

is closely related to the classic capillary number experimental curve. It is particularly noted that the data used in Chapter 4 of this volume is based on the (type 2) data in Table 5.1. Obviously, we can obtain the same result using the (type 1) data in the table for analysis study. However, the type 1 capillary number was defined with the "displacing velocity," and it is clearer to analyze the oil and water flow in the microscopic space with the "displacing velocity." The curve *QL* (II) is more suitable for the digital oil displacement research, and the parameters N_{cc}, N_{ct2}, N_{ct1} must be determined by the curve *QL* (II).

5.2 Digital study on the capillary number curve

A deeper understanding of the capillary number experimental curve requires to thoroughly figure out the difference between the classic capillary number experimental curve and the capillary number experimental curve *QL*, and to find out the cause of the difference.

Four curves were made in the capillary number experimental curve *QL* experiment under the condition above the limit capillary number N_{ct1}, and have the following two distinct characteristics.

Characteristic 1: The residual oil saturation value experiences a complex change with the increase of the capillary number. After the capillary number is higher than the limit capillary number N_{ct1}, the residual oil saturation increases relatively, and the relative maximum value is obtained under a certain capillary number. As the capillary number continuously increases, the corresponding residual oil saturation decreases. Chapter 4 clearly revealed the reasons for the special change of the residual oil saturation. In order to further study the problem, Fig. 5.3 shows a schematic diagram of the core microscopic oil-water distribution model given in Chapter 4, and a brief description is made. According to the analysis of the core microscopic oil-water distribution model, the displacement opens the oil-water coexistence space when the capillary number is higher than the limit capillary number N_{ct1}. This space has two distinct characteristics: (1) in the same pore size range $r_w \sim r_{WO}$, the "pure water" subspace coexists with the "pure oil" subspace and (2) the viscosity difference between oil and water results in that when the pore radius is smaller than r_w, the capillary number in starting the "residual oil" in the same pore diameter is higher than that in starting the "irreducible water." The two characteristics determine the special variation of the residual oil saturation.

Characteristic 2: The four curves are arranged in order. The table gives the "V/σ_{ow}" value for each experiment, wherein "V" is the "displacing velocity" defined by the formula (5.2). By organizing each group data and discarding the data with big deviations, the mean value of "V/σ_{ow}" is 2.44×10^{-3}, 3.86×10^{-3}, 1.14×10^{-2}, and 1.61×10^{-2} $(\text{mPa s})^{-1}$ in proper order, which are arranged from small to large. The figure shows that the corresponding experimental curves are arranged from top left to bottom right. Four different "V/σ_{ow}"

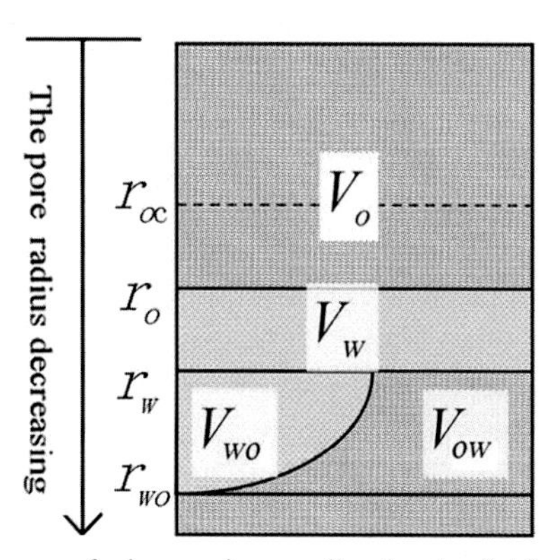

FIG. 5.3 Schematic diagram of microscopic core oil and water distribution.

values were employed to make the four curves, and the "V/σ_{ow}" value is the displacement condition of the experimental curve.

According to the above analysis and research, we had a clearer understanding of the capillary number experimental curve for displacement while the capillary number N_c is higher than the limit capillary number N_{ct1}. However, for the displacement while the capillary number N_c is lower than the limit capillary number N_{ct1}, the corresponding capillary number experimental curve needs to be studied further. The oil displacement process occurs in the pure oil space V_o with the pore radius higher than r_o, and in the pure water space V_w with the pore radius between $r_o \sim r_w$. There is no spatial structural feature of the oil-water coexistence space, and no flowing oil is trapped in the oil displacement, ensuring the monotonic change in the residual oil saturation of the experimental curve. However, in the analysis of the influence of the pore size change on the oil displacement effect, while the capillary number N_c is lower than the limit capillary number N_{cc}, the oil displacement occurs in the space V_{o1} with the pore radius higher than r_{oc} in the pure oil space V_o. Because of the large space pore radius, in spite of the difference in oil-water properties, such as viscosity, the capillary number in starting the "oil drop" in the pore is relatively small, which can be equal to the capillary number in starting the "water drop" in the same pore. Thus, no surfactant is needed in the displacing fluid. In the "V/σ_{ow}" value—the displacement condition—σ_{ow} is constant. The capillary number is increased only by the increase of the displacing velocity and the increase of the system viscosity in the oil displacement, and the displacement condition is unique, corresponding to a capillary number curve. When the oil displacement process enters the pure oil space V_{o2} with the pore radius lower than r_{oc}, due to the small space pore radius, the capillary number in starting the "oil drop" in the pore is relatively high, thus, the surfactant is needed in the displacing fluid, to reduce the influence of the capillary resistance, and further start the "oil drop" in the pore. The displacement condition—the "V/σ_{ow}" value—is not unique, corresponding to multiple capillary number curves.

Based on the above analysis, it is necessary to change the displacement conditions for the three sections of the capillary number in the capillary number curve, and further study the capillary number curve by using the digital displacement experiment research method.

The method used in the literature[1] was used to design a calculation model based on the experimental core condition: a two-dimensional strip model with three layers of equal thickness and different permeability in the longitudinal direction. The software IMCFS was used for calculation. Since the core microscopic oil-water distribution model has been constructed, the three oil layers of different permeability in the model can be treated as three-level core micro-models, and the combination of the three models is equivalent to the simulated experimental core micro-model. The main data of the model: the permeability of the three-level model is 100, 150, and $450 \times 10^{-3} \mu m^2$, respectively; the capillary number N_{cc}, N_{ct2}, and N_{ct1} are 1.0×10^{-4}, 2.5×10^{-3}, and 7.12×10^{-2}, respectively; the initial water saturation is 21%, 22%, and 24%, respectively, and distributed in the pure water space V_w and in the water subspace V_{wo} of the oil-water coexistence space. The oil saturation data of each oil space will be given in the analysis of the calculated results of the scheme. In particular, according to the research needs, the capillary numbers N_{cc}, N_{ct2}, and N_{ct1} are calculated with the data of the "type 2" curve in the previous section.

5.2.1 Displacement while the capillary number N_c is lower than the limit capillary number N_{cc}

While the capillary number N_c is lower than the limit capillary number N_{cc}, the displacement process occurs in the space V_{o1} with the pore radius higher than r_{oc} in the "pure oil space" V_o, and there is no need to add surfactant in the displacing fluid. The limit residual oil saturation S_{or}^L can be achieved only by increasing the displacing velocity. However, obtaining a relatively low residual oil saturation value through the simple water flooding requires a high flooding multiple. While a suitable amount of polymer can be added to the displacing fluid to increase the viscosity of the displacing fluid, and accelerate the oil phase flow rate. This can help reach the limit of the residual oil saturation by water flooding at a relatively low flooding multiple.

A group of oil displacement schemes were simulated and calculated. The scheme was terminated with 99.99% water content in the produced liquid. The results are shown in Table 5.2.

Scheme 1 in the table was the water flooding scheme, which was terminated when the water content of the produced liquid reached 99.98%. If the design requirements are not met, the scheme is terminated when the water content of the oil well produced liquid is 99.99%, because the oil displacement time exceeds the software's set time boundary of 100,000 days, the residual oil saturation of the scheme is 27.48%, and the flooding multiple is 42.08 PV at termination. Water Flooding Scheme 2 was calculated by increasing the flooding velocity by 0.5 times. The scheme ended normally when the flooding was 58.02 PV. The residual oil saturation of the scheme was 27.26%, the increase of the flooding velocity and the substantial increase of the flooding multiple reduced the residual oil saturation by 0.22%. Scheme 3 was calculated by doubling the flooding velocity of Scheme 1, the residual oil saturation of the scheme did not reduce, and the flooding multiple slightly decreased. Based on Scheme 1, the polymer solution with concentration of 100 mg/L was injected to calculate Scheme 4, and the solution subsurface viscosity was 1.25 mPa s. Compared with the water flooding scheme, the residual oil saturation slightly reduced, and the flooding multiple greatly decreased. Scheme 4 was calculated by gradually increasing the polymer concentration in the scheme system to 3500 mg/L. As the subsurface viscosity gradually increased, the capillary number in the oil displacement process gradually increased, while the residual oil saturation gradually decreased. In Scheme 8, the polymer concentration was 3500 mg/L,

TABLE 5.2 Scheme calculation results of water flooding and polymer flooding.

Case no.	Liquid injection rate (m^3/d)	Polymer concentration (mg/L)	Total injection multiple (PV)	Produced liquid watercut F_w (%)	Recovery factor R (%)	Residual oil saturation (%)	Displacement velocity V (10^{-7} m/s)	Oil-water interfacial tension σ_{ow} (mN/m)	Water viscosity μ_w (mPa s)	Capillary number N_c (10^{-8})
1	4.32	0	42.08	99.98	64.62	27.48	7.80	22.9	0.6	2.04
2	6.48	0	58.02	99.99	64.90	27.26	11.7	22.9	0.6	3.07
3	8.64	0	57.98	99.99	64.90	27.26	15.6	22.9	0.6	4.09
4	4.32	100	40.56	99.99	65.14	27.07	7.80	22.9	1.25	4.26
5	4.32	1000	13.34	99.99	65.50	26.79	7.80	22.9	10.09	34.4
6	4.32	2000	6.25	99.99	65.59	26.72	7.80	22.9	25.75	87.7
7	4.32	3000	3.43	99.99	65.62	26.70	7.80	22.9	50.64	172
7*	4.32	3340	3.09	99.99	65.63	26.69	7.80	22.9	61.88	211
8	4.32	3500	2.88	99.99	65.63	26.69	7.80	22.9	71.35	243
9	4.32	4000	2.56	99.99	65.63	26.69	7.80	22.9	89.09	303
10	6.48	3500	2.90	99.99	65.63	26.69	11.7	22.9	70.17	359

and the residual oil saturation was 26.69%. Scheme 9 was calculated by keeping the flooding velocity unchanged, and increasing the polymer concentration to 4000 mg/L. Then Scheme 10 was calculated by increasing the flooding velocity by 0.5 times, and keeping the polymer concentration at 3500 mg/L. The residual oil saturation at the termination of the two schemes was 26.69%, which was the same as that of Scheme 8. Therefore, it was determined that the limit of the water flooding residual oil saturation S_{or}^{L} of the model is 26.69%. Scheme 7* was calculated in encrypted way, to determine that the residual oil saturation reaches the limit value S_{or}^{L} when the capillary number is 2.11×10^{-6}. The limit capillary number N_{cc} should be the capillary number where the residual oil saturation has a transitional change, and it should be higher than the capillary number 3.59×10^{-6} in Scheme 10.

Looking back at the experiment on the research of the capillary number experimental curve *QL*, it is found that the capillary number N_{cc} is determined under the ASP flooding. Therefore, the capillary number N_{cc} is also determined in the subsequent ASP flooding study.

5.2.2 Displacement while the capillary number N_c is between the limit capillary number N_{cc} and N_{ct1}

While the capillary number is between the limit capillary number N_{cc} and N_{ct1}, the displacement process gradually enters the space V_{o2} with the pore radius less than r_{oc} in the "pure oil space" V_o. Due to the relative decrease of the pore size, the surfactant must be added to the displacing fluid. The low interfacial tension of the system overcomes the capillary resistance, and the seepage velocity of the system is the driving force. Coupled with the viscosity of the system, each of the three plays an indispensable role in the oil displacement process. The displacement condition is quite similar to that of the "pure oil" subspace in the oil-water coexistence space. Both of them are the "pure oil" space, and as the capillary number increases in displacement, the displaced pore size corresponds to a smaller space. The research in the literature[1] shows that using the simulation calculation research method, it is obtained that "there are multiple capillary number curves for displacement while the capillary number is higher than the limit capillary number N_{ct}," and the calculated results are verified by experiment. The simulation study method was also used to study the displacement while the capillary number is between the limit capillary number N_{cc} and N_{ct1}. The five groups of schemes were calculated by changing V/σ_{ow} value, and the results are listed in Table 5.3.

First, we studied the first group of data calculation schemes in Table 5.3. This was designed after re-studying the four pairs of experiments while the capillary number is between N_{cc} and N_{ct2} in the capillary number experimental curve *QL*—Nos. 3–10 experiment in Table 5.1. Four experimental points of the curve were determined for the four pairs of oil displacement experiment. The average values of V/σ_{ow} between the seepage velocity V and the interfacial tension σ_{ow} of the four experimental points were 1.44×10^{-5}, 1.50×10^{-4},

TABLE 5.3 The calculation results of ASP flooding scheme (1).

Curve	Index	Scheme									
		1	2	*3*	*4*	*5*	*6*	*7*	*8*	*9*	*10*
1	V, 10^{-6} m/s	1.56	1.56	1.56	1.56	1.56	1.56	1.56	1.56	2.34	2.34
	σ_{ow}, mN/m	0.05	0.05	0.05	0.05	0.05	0.025	0.015	0.015	0.015	0.015
	μ_w, mPas	8.68	8.82	35.30	50.73	71.34	89.11	71.39	89.16	70.23	87.69
	V/σ_{ow}, 10^{-5} $(\text{mPas})^{-1}$	3.12	3.12	3.12	3.12	3.12	6.24	10.4	10.4	1.56	1.56
	N_c, 10^{-4}	2.71	2.75	11.0	15.8	22.3	55.6	74.2	92.7	110	137
	S_{or},%	26.69	26.68	24.41	22.51	21.53	18.92	18.33	18.08	18.06	18.06
2	V, 10^{-6} m/s	1.56	1.56	1.56	1.56	1.56	1.56	1.56	1.56	1.56	1.56
	σ_{ow}, mN/m	0.01	0.01	0.01	0.01	0.01	0.01	0.01	0.01	0.01	0.01
	μ_w, mPas	4.33	7.57	10.21	16.14	25.75	35.08	41.56	50.67	71.40	89.16
	V/σ_{ow}, 10^{-4} $(\text{mPas})^{-1}$	1.56	1.56	1.56	1.56	1.56	1.56	1.56	1.56	1.56	1.56
	N_c, 10^{-4}	6.75	11.8	15.9	25.2	40.2	54.7	64.8	79.0	111	139
	S_{or}, %	26.30	25.49	24.77	22.65	20.91	19.48	18.75	18.23	18.06	18.06
3	V, 10^{-7} m/s	7.80	7.80	7.80	7.80	7.80	7.80	7.80	7.80	7.80	7.80
	σ_{ow}, 10^{-1} mN/m	0.05	0.05	0.05	0.05	0.05	0.05	0.05	0.05	0.05	0.05
	μ_w, mPas	4.43	7.75	10.47	16.57	26.46	36.05	42.71	52.08	73.39	91.64
	V/σ_{ow}, 10^{-4} $(\text{mPas})^{-1}$	1.56	1.56	1.56	1.56	1.56	1.56	1.56	1.56	1.56	1.56

	N_c, 10^{-4}	6.91	12.1	16.3	25.8	41.3	56.2	66.6	81.2	114	143
	S_{or}, %	26.27	25.44	24.71	22.54	20.84	19.38	18.69	18.20	18.06	18.06
4	V, 10^{-6} m/s	1.56	1.56	1.56	1.56	1.56	1.56	1.56	1.56	1.56	1.56
	σ_{ow}, 10^{-2} mN/m	0.25	0.25	0.25	0.25	0.25	0.25	0.25	0.25	0.25	0.25
	μ_w,mP·s	1.26	2.42	4.29	7.45	10.09	16.13	25.77	50.73	71.47	83.32
	V/σ_{ow}, 10^{-4} $(\text{mPa s})^{-1}$	6.24	6.24	6.24	6.24	6.24	6.24	6.24	6.24	6.24	6.24
	N_c, 10^{-4}	7.86	15.1	26.8	46.5	63.0	101	161	317	446	520
	S_{or},%	26.58	25.36	23.51	21.58	20.73	18.75	18.13	18.07	18.06	18.17
5	V, 10^{-6} m/s	3.12	3.12	3.12	3.12	3.12	3.12	3.12	3.12	3.12	3.12
	σ_{ow}, 10^{-1} mN/m	0.05	0.05	0.05	0.05	0.05	0.05	0.05	0.05	0.05	0.05
	μ_w, mP·s	1.24	2.37	4.18	7.25	9.82	15.69	25.06	49.30	69.46	80.97
	V/σ_{ow}, 10^{-4} $(\text{mPa s})^{-1}$	6.24	6.24	6.24	6.24	6.24	6.24	6.24	6.24	6.24	6.24
	N_c, 10^{-4}	7.74	14.8	26.1	45.2	61.3	97.9	156	308	433	505
	S_{or}, %	26.61	25.41	23.61	21.64	20.81	18.82	18.13	18.07	18.06	18.17

1.06×10^{-4}, and 6.26×10^{-3} $(\text{mPa s})^{-1}$ respectively. According to the experimental results, two key points on the curve were determined. For Experimental Point 1, it was determined that the limit capillary number N_{cc} was about 8.42×10^{-5}, and the corresponding residual oil saturation S_{or}^{L} was about 27.3%. For Experimental Point 4, it was determined the limit capillary number N_{ct2} was about 6.26×10^{-3}, and the corresponding residual oil saturation S_{or}^{H} was about 18.7%. Inspired by the design of the core flooding experiment, the group of oil displacement schemes were designed with four "V/σ_{ow}" values, and the corresponding polymer concentration was configured for each scheme, to ensure that the capillary number of the scheme in the oil displacement process increases sequentially and the residual oil saturation gradually decreases. For Scheme 1 in the table, the capillary number was 2.71×10^{-4}, and the corresponding residual oil saturation value was 26.69%, which is equivalent to the limit of the residual oil saturation by water flooding given in the previous section. For Scheme 2, the capillary number was 2.75×10^{-4}, and the corresponding residual oil saturation value was 26.68%, which can be considered as the "significant decrease in the residual oil saturation," thus, we determined that the limit capillary number N_{cc} was 2.71×10^{-4}, and the limit residual oil saturation was 26.69%. Subsequently, the capillary number of the scheme system gradually increased, all of them were in displacement in the condition above the limit capillary number N_{cc}, and the residual oil saturation gradually decreased. Scheme 9 and Scheme 10 had the same residual oil saturation value of 18.06%. The capillary number of Scheme 9 was 1.10×10^{-2}, from which we can determine that the limit capillary number N_{ct2} is not higher than 1.10×10^{-2}.

For each scheme of Group 2, the seepage velocity V was 1.56×10^{-6} m/s, the interfacial tension of the system σ_{ow} was 0.01 mN/m, the V/σ_{ow} value was 1.56×10^{-4} $(\text{mPa s})^{-1}$. The system viscosity gradually increased, the capillary number N_c gradually increased, and the corresponding residual oil saturation gradually decreased. For Scheme 1, the capillary number N_c was 6.75×10^{-4}, and the residual oil saturation was 26.3%. The displacement in each scheme occurred while the capillary number is higher than the limit capillary number N_{cc}. The residual oil saturation of Schemes 9 and 10 was 18.06%, and the capillary number N_c of Scheme 9 was 1.11×10^{-2}. For each scheme of Group 3, the seepage velocity V was 7.80×10^{-7} m/s, the interfacial tension of the system σ_{ow} was 0.005 mN/m, the V/σ_{ow} value was 1.56×10^{-4} $(\text{mPa s})^{-1}$. The system viscosity gradually increased, the capillary number N_c gradually increased, and the corresponding residual oil saturation gradually decreased. For Scheme 1, the capillary number N_c was 6.91×10^{-4}, and the residual oil saturation was 26.27%. The residual oil saturation of Schemes 9 and 10 was 18.06%, and the capillary number N_c of Scheme 9 was 1.14×10^{-2}. In the two groups of schemes, the polymer concentration in the scheme system with the same number was the same. Schemes in Group 3 had low seepage velocity, low shear thinning, higher viscosity retention rate, slightly larger capillary number value and slightly lower residual oil saturation value.

For schemes in Groups 4 and 5, the V/σ_{ow} value was 6.24×10^{-4} (mPa s)$^{-1}$. The setting for the seepage velocity and the interfacial tension of the two groups of schemes were similar to those of the first two groups of schemes. The oil displacement schemes were calculated under the same polymer concentration. Schemes of Group 4 had low seepage velocity V, slightly higher subsurface viscosity, and slightly lower residual oil saturation. The residual oil saturation of the two Schemes 1 was 26.58% and 26.61% respectively, lower than the limit of the residual oil saturation by water flooding 26.69%, and the capillary number was 7.86×10^{-4} and 7.74×10^{-4} respectively, higher than the limit capillary number N_{cc} 2.79×10^{-4}. The capillary number of Scheme 9 in Groups 4 and 5 was 4.46×10^{-2} and 4.33×10^{-2} respectively, and the residual oil saturation of the two schemes was 18.06%. The capillary number of Scheme 10 in Groups 4 and 5 was 5.20×10^{-2} and 5.05×10^{-2} respectively, and the two Scheme 10 shared the equal residual oil saturation 18.17%, which was obviously higher than the residual oil saturation value of the two Scheme 9. The two Scheme 10 entered the "type II" displacement while the capillary number is higher than the limit capillary number N_{ct1}. The limit capillary number N_{ct1} should be greater than 4.46×10^{-2} but less than 5.05×10^{-2}.

While the capillary number is between the limit capillary number N_{cc} and N_{ct1}, the displacement process entered the "pure water space" V_w. The capillary number increased, the pore size in the open space became smaller, but no oil was produced in the new open space. The oil output did not increase and the residual oil saturation remained unchanged.

The scheme results of Group 1 are listed in Table 5.4, most of which were extracted from Table 5.3, and some from the scheme in Table 5.2 and the additional scheme. For the in-depth study, the residual oil saturation content for the three-level micro-space of the calculation model was added to the scheme information.

By analyzing Schemes 6–9, the capillary number increased in turn, and their residual oil saturation value was 18.06%, which was determined to be the limit residual oil saturation S_{or}^H, close to the average value 18.76% of the residual oil saturation S_{or}^H obtained by the oil displacement experiment. This value is the volume percentage v_{ow} of the oil content of the "pure oil" subspace V_{ow} in the micro-space of the model. By analyzing the residual oil saturation data of the three-level micro-space of schemes in Group 4, the first-level spatial value was 15.77%, the second-level spatial value was close to 18.38%, and the third-level spatial value was about 20.07%. The residual oil value in the three-level micro-space is the volume percentage v_{ow} of the oil content in the "pure oil" subspace V_{ow} in the oil-water coexistence space in the three-level microscopic model. Schemes 5 and 6 in Table 5.4 were Schemes 8 and 9 in Table 5.3, respectively. For Scheme 5, the residual oil saturation was 18.08%, higher than the limit residual oil saturation S_{or}^H, and the capillary number 9.27×10^{-1} should be lower than the limit capillary number N_{ct2}. For Scheme 6, the residual oil saturation was 18.06%, and the capillary number

TABLE 5.4 Calculated results of ASP flooding scheme (2).

Scheme no.	Oil-water interfacial tension σ_{ow} (mN/m)	Water viscosity μ_w (mPa s)	Seepage velocity V (10^{-7} m/s)	Capillary number N_c (10^{-4})	Residual oil saturation (%)	Remaining oil saturation (%)		
						Tertiary microscopic space (low permeability)	*Second order microscopic space (medium permeability)*	*First order microscopic space (high permeability)*
1	22.9	61.88	7.8	**0.021**	**26.69**	**28.04**	**27.03**	**25.01**
2	22.9	70.17	11.7	0.021	26.69	28.04	27.03	25.01
3	**0.05**	**8.68**	**15.6**	**2.71**	**26.69**	**28.49**	**27.36**	**24.27**
4	0.05	8.82	15.6	2.75	26.68	28.48	27.36	24.25
5	0.015	89.16	15.6	92.7	18.08	20.13	18.38	15.77
5*	0.015	62.95	23.4	98.2	18.07	20.08	18.39	15.77
6	**0.015**	**70.23**	**23.4**	**110**	**18.06**	**20.07**	**18.38**	**15.77**
7	0.05	73.39	7.8	111	**18.06**	20.07	18.38	15.77
8	0.01	71.40	15.6	114	**18.06**	20.07	18.38	15.77
9	**0.0025**	**71.47**	**15.6**	**446**	**18.06**	**20.07**	**18.38**	**15.77**
9*	0.005	49.65	46.8	465	18.07	20.08	18.38	15.77
10	0.005	80.97	31.2	505	18.17	20.07	18.34	16.14

The bold values in Schemes 1, 3, 6, and 9 are corresponding respectively to the start point, N_{cc}, N_{ct2}, and N_{ct1} in the capillary number experimental curve *QL*.

1.10×10^{-2} should not be lower than the limit capillary number N_{ct2}. Scheme 5* was calculated in encrypted way, the residual oil saturation was 18.07%, the capillary number was 9.82×10^{-1}, and we can determine that the limit capillary number N_{ct2} of the model is 1.10×10^{-2}. Scheme 9 in Table 5.4 was Scheme 9 of Group 4 in Table 5.3, and Scheme 10 was Scheme 10 of Group 5 in Table 5.3. Scheme 9* was calculated in encrypted way, the residual oil saturation was 18.07%, and we can determine that the limit capillary number N_{ct1} is 4.65×10^{-2}.

At this point, relevant parameter data of the three-level microcosmic model can be supplemented. As summarized in Chapter 4, the initial core "irreducible water" saturation S_{wr}^L is distributed in the "pure water" space V_w and the "irreducible water" subspace V_{wo}, the initial oil saturation S_o is equal to $1 - S_{wr}^L = v_{o1} + v_{o2} + v_{ow}$, wherein, v_{o1}, v_{o2}, v_{ow} are the pore volume percentage parameter of the oil content in the "pure oil" subspace V_{o1}, V_{o2}, V_{ow} respectively. The limit capillary number N_{cc} of water flooding corresponds to the amount of oil in the subspace V_{o1}, and the limit residual oil saturation S_{or}^L is equal to $S_o - v_{o1}$. Then, the capillary number N_c of ASP flooding increases to the limit capillary number N_{ct2}, which corresponds to the amount of oil in the subspace V_{o2}, the limit residual oil saturation is S_{or}^H, and the residual oil is in the "pure oil" subspace V_{ow}, $S_{or}^H = v_{ow}$. After confirming the relevant value of S_o, S_{or}^L, S_{or}^H, we can calculate $v_{o1} = S_{or}^L - S_{or}^H$, $v_{o2} = S_{or}^L - S_{or}^H$. As mentioned, the initial water saturation S_{wr}^L of the three-level microscopic model is 21%, 22%, and 24% respectively, from which we can calculate the initial oil saturation S_o of the three-level microscopic model is 79%, 78%, and 76% respectively. Scheme 1 in Table 5.4 was Scheme 7* in Table 5.2, and the capillary number was 2.11×10^{-6}. In Table 5.2, Schemes 8–10 have the same residual oil saturation of 26.69%. For comparison, Scheme 10 in the table was listed as Scheme 2 in Table 5.4, and Schemes 3 and 4 in Table 5.4 were Schemes 1 and 2 of Group 1 in Table 5.3. In Table 5.4, the residual oil saturation of Scheme 3 was still 26.69%, and that of Scheme 4 was 26.68%, indicating a significant change in the residual oil saturation. Thus, we can confirm that the limit capillary number N_{cc} of the model was 2.71×10^{-4}, the limit residual oil saturation S_{or}^L of the model was 26.69%, the corresponding values of the three-level submodel were 24.27%, 27.36%, and 28.49% respectively. The capillary number 1.10×10^{-2} of Scheme 6 was the limit capillary number N_{ct2}, the limit residual oil saturation S_{or}^H of the model was 18.06%, the volume percentage v_{ow} of the oil content in the oil-water coexistence space of the three-level microcosmic model was 15.77%, 18.38%, and 20.07% respectively. The volume percentage v_{o1} of the oil content in the "pure oil" space V_{o1} was 50.97%, the corresponding values of the three-level submodel were 54.73%, 50.64%, and 47.51% respectively. The volume percentage v_{o2} of the oil content in the "pure oil" space V_{o2} was 8.63%, the corresponding values of the three-level submodel were 8.50%, 8.98%, and 8.42% respectively.

5.2.3 Displacement while the capillary number N_c is higher than the limit capillary number N_{ct1}

Based on the in-depth study on the condition that the capillary number is lower than the limit capillary number N_{ct1}, it is necessary to further study on the condition that the capillary number is higher than the limit capillary number N_{ct1}, to make a relatively complete "capillary number 'digitized' oil displacement experimental curve."

The different values of V/σ_{ow} were used to calculate the corresponding curve, and the curve distribution law was studied in detail. Table 5.5 shows the calculation results of six groups of schemes.

The V/σ_{ow} value of schemes in Groups 1 and 2 was 9.36×10^{-3} $(\text{mPa s})^{-1}$. The seepage velocity of schemes of Group 1 was 9.36×10^{-6} m/s. For Scheme 1, the capillary number in oil displacement was 6.51×10^{-2}, slightly higher than the limit capillary number N_{ct1} 4.46×10^{-2}, and the residual oil saturation was 18.3%, higher than the limit residual oil saturation 18.06%. As the viscosity of the scheme system increased, the capillary number increased, and the residual oil saturation also increased. For Scheme 7, the residual oil saturation reached the maximum 20.61%, the corresponding capillary number was 0.361, the system viscosity further increased, and the residual oil saturation began to decrease. For Scheme 10, the capillary number was 0.798, and the residual oil saturation was 19.36%. The seepage velocity of schemes of Group 2 was 2.34×10^{-6} m/s. For Scheme 1, the capillary number in oil displacement was 6.87×10^{-2}, and the residual oil saturation was 18.33%. In Scheme 7, the residual oil saturation reached the maximum value 20.4%, and the capillary number was 0.382. For Scheme 10, the capillary number was 0.819, and the residual oil saturation was 18.75%. The two groups of scheme systems with the same number had the same polymer concentration, and the corresponding values of the capillary number and the residual oil saturation differed greatly. This is the result of the difference in seepage velocity. The seepage velocity of schemes of Group 1 was four times that of Group 2.

The V/σ_{ow} value of schemes in Groups 3 and 4 was 3.744×10^{-2} $(\text{mPa s})^{-1}$. The seepage velocity of schemes of Group 3 was 9.36×10^{-6} m/s. For Scheme 1, the capillary number in oil displacement was 8.50×10^{-2}, and the residual oil saturation was 18.67%. For Scheme 5, the residual oil saturation reached the maximum 22.72%, and the corresponding capillary number was 0.56. For Scheme 10, the capillary number was 2.54, and the residual oil saturation was 18.91%. The seepage velocity of schemes of Group 4 was 7.488×10^{-7} m/s. For Scheme 1, the capillary number in oil displacement was 9.17×10^{-2}, and the residual oil saturation was 19%. For Scheme 5, the residual oil saturation reached the maximum 22.86%, and the corresponding capillary number was 0.621. For Scheme 10, the capillary number was 2.75, and the residual oil saturation was 18.58%. In the two groups of schemes with the same number, the corresponding values of the capillary number and the residual oil saturation differed greatly, and the former was 12.5 times of the latter.

TABLE 5.5 Calculated results of ASP flooding scheme (3).

Curve	Index	Scheme									
		1	*2*	*3*	*4*	*5*	*6*	*7*	*8*	*9*	*10*
1	V, 10^{-6} m/s	9.36	9.36	9.36	9.36	9.36	9.36	9.36	9.36	9.36	9.36
	σ_{ow}, mN/m	0.001	0.001	0.001	0.001	0.001	0.001	0.001	0.001	0.001	0.001
	μ_w, mPa s	6.95	7.96	9.37	14.97	23.93	32.56	38.55	47.67	67.73	85.27
	V/σ_{ow}, 10^{-3} (mPa s)$^{-1}$	9.36	9.36	9.36	9.36	9.36	9.36	9.36	9.36	9.36	9.36
	N_c, 10^{-2}	6.51	7.45	8.79	14.0	22.4	30.5	36.1	44.6	63.4	79.8
	S_{or}, %	18.30	18.34	18.62	19.62	19.51	20.37	20.61	20.39	19.77	19.36
2	V, 10^{-6} m/s	2.34	2.34	2.34	2.34	2.34	2.34	2.34	2.34	2.34	2.34
	σ_{ow},10^{-2} mN/m	0.025	0.025	0.025	0.025	0.025	0.025	0.025	0.025	0.025	0.025
	μ_w, mPa s	7.34	8.42	9.93	15.86	25.37	34.54	40.76	49.73	69.32	87.49
	V/σ_{ow}, 10^{-3} (mPa s)$^{-1}$	9.36	9.36	9.36	9.36	9.36	9.36	9.36	9.36	9.36	9.36
	N_c, 10^{-2}	6.87	7.88	9.29	14.8	23.7	32.3	38.2	46.5	64.9	81.9
	S_{or}, %	18.33	18.34	19.77	19.45	19.60	20.26	20.40	20.09	19.38	18.75
3	V, 10^{-6} m/s	9.36	9.36	9.36	9.36	9.36	9.36	9.36	9.36	9.36	9.36
	σ_{ow}, 10^{-3} mN/m	0.25	0.25	0.25	0.25	0.25	0.25	0.25	0.25	0.25	0.25
	μ_w, mPa s	2.27	3.98	6.95	9.39	14.96	23.64	27.93	32.26	47.12	67.85
	V/σ_{ow}, 10^{-2} (mPa s)$^{-1}$	3.744	3.744	3.744	3.744	3.744	3.744	3.744	3.744	3.744	3.744
	N_c, 10^{-2}	8.50	14.9	26.0	35.2	56.0	88.5	105	121	176	254
	S_{or}, %	18.67	20.21	19.77	21.29	22.72	22.09	21.41	20.74	19.67	18.91

Continued

TABLE 5.5 Calculated results of ASP flooding scheme (3)—cont'd

Curve	Index	Scheme									
		1	2	*3*	*4*	5	6	*7*	*8*	*9*	*10*
4	V, 10^{-7} m/s	7.488	7.488	7.488	7.488	7.488	7.488	7.488	7.488	7.488	7.488
	σ_{ow}, 10^{-5} mN/m	2.0	2.0	2.0	2.0	2.0	2.0	2.0	2.0	2.0	2.0
	μ_w, mPs	2.45	4.37	7.67	10.38	16.58	26.19	30.90	35.73	52.15	73.41
	V/σ_{ow}, 10^{-2} (mPas)$^{-1}$	3.744	3.744	3.744	3.744	3.744	3.744	3.744	3.744	3.744	3.744
	N_c, 10^{-2}	9.17	16.4	28.7	38.9	62.1	98.1	116	134	195	275
	S_{or}, %	19.00	20.16	19.64	21.49	22.86	22.36	21.78	21.05	19.50	18.58
5	V, 10^{-6} m/s	1.872	1.872	1.872	1.872	1.872	1.872	1.872	1.872	1.872	1.872
	σ_{ow}, 10^{-2} mN/m	0.002	0.002	0.002	0.002	0.002	0.002	0.002	0.002	0.002	0.002
	μ_w, mPas	2.39	7.39	10.01	16.0	25.47	34.82	41.24	50.27	70.77	88.38
	V/σ_{ow}, 10^{-2} (mPas)$^{-1}$	9.36	9.36	9.36	9.36	9.36	9.36	9.36	9.36	9.36	9.36
	N_c, 10^{-1}	2.24	6.92	9.37	15.0	23.8	32.6	38.6	47.1	66.2	82.7
	S_{or}, %	20.15	23.17	23.18	22.68	21.38	19.99	19.41	18.84	18.07	17.53
6	V, 10^{-6} m/s	2.34	2.34	2.34	2.34	2.34	2.34	2.34	2.34	2.34	2.34
	σ_{ow},10^{-3} mN/m	0.025	0.025	0.025	0.025	0.025	0.025	0.025	0.025	0.025	0.025
	μ_w, mPas	2.37	7.33	9.92	15.85	25.15	34.50	40.86	49.82	70.13	87.58
	V/σ_{ow}, 10^{-2} (mPas)$^{-1}$	9.36	9.36	9.36	9.36	9.36	9.36	9.36	9.36	9.36	9.36
	N_c, 10^{-1}	2.22	6.86	9.29	14.8	23.5	32.3	38.2	46.6	65.6	82.0
	S_{or}, %	20.02	23.16	23.21	22.68	21.44	20.00	19.46	18.88	18.12	17.60

The V/σ_{ow} value of schemes in Groups 5 and 6 was 9.36×10^{-2} $(\text{mPa s})^{-1}$. The seepage velocity of schemes of Group 5 was 1.872×10^{-6} m/s. For Scheme 1, the capillary number in oil displacement was 0.224, and the residual oil saturation was 20.15%. For Scheme 3, the residual oil saturation reached the relative maximum 23.18%, and the corresponding capillary number was 0.937. For Scheme 10, the capillary number was 8.27, and the residual oil saturation was 17.53%, lower than the limit residual oil saturation 18.06%. The seepage velocity of schemes of Group 6 was 2.34×10^{-6} m/s. For Scheme 1, the capillary number in oil displacement was 0.222, and the residual oil saturation was 20.02%. For Scheme 3, the residual oil saturation reached the maximum 23.21%, and the corresponding capillary number was 0.929. For Scheme 10, the capillary number was 8.20, and the residual oil saturation was 17.60%, lower than the limit residual oil saturation. In the two groups of schemes with the same number, the corresponding values of the capillary number and the residual oil saturation were slightly different, which was the result of similar seepage velocity between the two groups of schemes.

5.2.4 Capillary number "digitized" experimental curve

According to the data of Table 5.2, Table 5.3, and Table 5.5, the capillary number "digitized" experimental curve was plotted, as shown in Fig. 5.4, which is a relatively more complete capillary number curve. The curve was divided into three parts with the limit capillary number N_{cc}, N_{ct1} as the dividing point.

While the capillary number N_c is lower than the limit capillary number N_{cc}, the displacement process occurs in the space V_{o1} of the core microcosmic pure oil space V_o. With relatively high pore radius, there is no need to add the

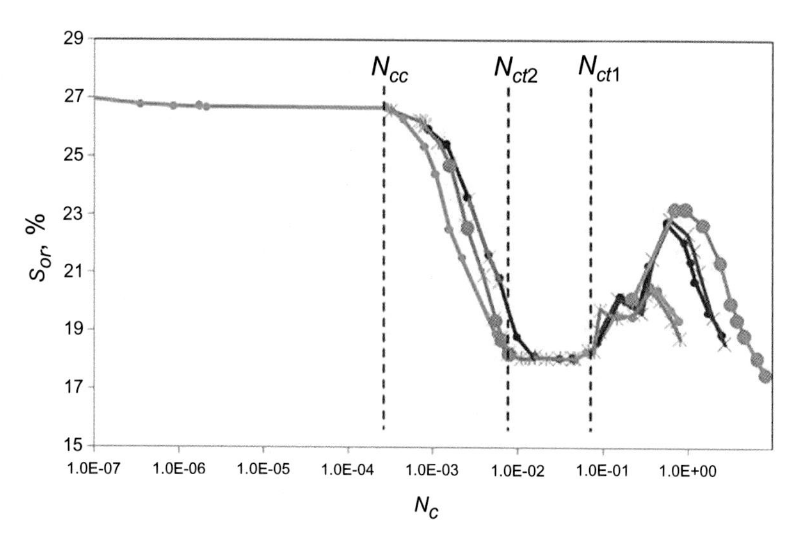

FIG. 5.4 Capillary number "digitization" experimental curve.

surfactant to reduce the interfacial tension in the oil displacement system. Only a certain displacing velocity and a suitable system viscosity can empty the crude oil stored in the microcosmic pure oil space V_{o1}. A capillary number curve was obtained against the displacement under a defined displacement condition. While the capillary number N_c is between the limit capillary number N_{cc} and N_{ct1}, the oil displacement process is expanded to the whole core microcosmic pure oil space V_o and the pure water space V_w. The pore size shrinks, the capillary resistance increases in displacement, and it is necessary to add the surfactant to the displacing fluid. The system of different interfacial tensions σ_{ow} and the given velocity V determined the "displacement conditions" of different V/σ_{ow} values. Under different displacement conditions, a cluster of curves were obtained. They were arranged orderly starting from the point (N_{cc}, S_{or}^L). From the second curve, each curve corresponds to a defined V/σ_{ow} value, which gradually increases from left to right. The first line is the "envelope line" of this cluster of curves. Due to the shear thinning effect of the seepage velocity on the viscosity of the system, the coincidence degree of two curves with the same V/σ_{ow} value is affected by the difference of the seepage velocity. With the large seepage velocity difference, two curves have a poor coincidence degree. For ASP flooding while the capillary number N_c is higher than the limit capillary number N_{ct}, the displacement process enters the oil-water coexistence space with the pore radius smaller than r_w, where the pure oil subspace V_{ow} and the pure water subspace V_{wo} exist in parallel. Under different V/σ_{ow} values, a cluster of capillary number curves were obtained in the experiment. Starting from the vicinity of the point (N_{cc}, S_{or}^L), the curve gradually increases with the capillary number N_c, the residual oil saturation S_{or} gradually increases to a relative maximum value, and then gradually decreases; the V/σ_{ow} value gradually increases, and the curve is distributed regularly. With the small V/σ_{ow} value, the curve is at the lower left. Apparently, in the case of a large difference in the seepage velocity, the two curves with the same V/σ_{ow} value are obviously separated. The "envelope line" of the cluster of curves plotted in the figure was the curve plotted according to the five groups of data in the table. The V/σ_{ow} value of this group of results is sufficiently large, so that this curve can also be approximated as the "envelope line" of the ASP flooding capillary number curve when the capillary number N_c is higher than the limit capillary numberN_{ct1}.

5.2.5 Mathematical description of the capillary number experimental curve *QL*

In the process of water flooding or polymer solution displacement, the capillary number N_c is lower than the limit capillary number N_{cc}, and the core oil-water interfacial tension σ_{ow} is constant B. Coupled with the displacing velocity V, the water viscosity μ_w, and the residual oil saturation S_{or}, we can obtain the following binary composite function:

$$S_{or} = F(N_c) = F(\mu_w, V/B) \tag{5.4}$$

The function curve shows a gentle decline.

For ASP flooding while the capillary number is between the limit capillary number N_{cc} and N_{ct1}, the core oil-water interfacial tension σ_{ow}, the displacing velocity V, the water viscosity μ_w, and the residual oil saturation S_{or} can be expressed as a ternary composite function:

$$S_{or} = F(N_c) = F(\mu_w, V/\sigma_{ow}) \tag{5.5}$$

Starting from the point (N_{cc}, S_{or}^L), the function curve gradually increases with the capillary number N_c, the residual oil saturation S_{or} rapidly decreases first, reaches the maximum value S_{or}^H at the capillary number of N_{ct}, and then presents a flat change until reaching to the limit capillary number N_{ct1}. This is a multi-value function, that is, the function curve is a cluster of curves, which have a common starting point (N_{cc}, S_{or}^L) and a common ending point (N_{ct1}, S_{or}^H), where the abscissa value N_{ct} of a point (N_{ct}, S_{or}^H) on the unique curve is equal to the limit capillary number N_{ct2}. The abscissa value of the point (N_{ct}, S_{or}^H) on the other curves is in the range of (N_{ct2}, N_{ct1}). We can further write the following formula.

$$S_{or} = F(N_c) = \begin{cases} F_l(\mu_w, V/\mu_{ow}) \\ F_m(C_m \cdot \mu_w) \end{cases} \tag{5.6}$$

Wherein, $m = 2, 3, 4\ldots$, the constant $C_m = V/\sigma_{ow}$ for the appropriate seepage velocity V and the system interfacial tensionσ_{ow}.

For ASP flooding while the capillary number is higher than the limit capillary number N_{ct1}, the core oil-water interfacial tension σ_{ow}, the displacing velocity V, the water viscosity μ_w, the proper constant C_m ($C_m = V/\sigma_{ow,}\ m = 1, 2, 3\ldots$), and the residual oil saturation S_{or} can be expressed as the following unary composite function:

$$S_{or} = F_m(N_c) = F(C_m \cdot \mu_w) \tag{5.7}$$

Each function curve gradually increases with the viscosity of the system, the residual oil saturation S_{or} first increases, and gradually decreases after reaching the maximum. As C_m increases, the corresponding curve position moves to the upper right.

5.2.6 Deep understanding of the capillary number experimental curve

The digital study of the capillary number experimental curve further deepens the understanding of capillary number experimental curve.

The experimental core is the objective experimental basis. A capillary number curve corresponding to the residual oil saturation S_{or} can be obtained through a group of experiments under the determined displacement conditions. The V/σ_{ow} value is the element of the ASP flooding condition. A group of experiments can have a V/σ_{ow} value, or a number of V/σ_{ow} values. The trend of the curve is determined by the V/σ_{ow} value of the ASP flooding experiment starting from the limit capillary number N_{cc}. The classical capillary number curve is also a group of experimental curves obtained under a defined displacement condition, and the relatively large V/σ_{ow} value is taken in the initial experiment of ASP flooding. We obtained the limit residual oil saturation S_{or}^H and the corresponding limit capillary number N_{ct}. The scholars named "N_{ct}" as the "capillary number in which the residual oil saturation no longer decreases." Apparently, the scholars have made many displacement experiments with higher capillary number after making the limit residual oil saturation S_{or}^H. The residual oil saturation of these experiments is equal to or higher than the limit residual oil saturation S_{or}^H. University of Texas scholars made the increase and decrease of the residual oil saturation for displacement while the capillary number is higher than the limit capillary number N_{ct}, which was seen in the results calculated using the UTCHEM software, shown in Chapter 4. However, they do not get the case in which the residual oil saturation S_{or} is lower than the limit residual oil saturation S_{or}^H.

The digital study of the capillary number curve deepens the understanding of the core oil-water distribution parameters and key parameters of the capillary number curve. Oil-water distribution in core space: the irreducible water exists in the pure water space V_w and the water subspace V_{wo} of the oil-water coexistence space, and the saturation value is S_{wr}^L; the core initial oil saturation is $1-S_{wr}^L$, distributed in the pure oil space V_o and the pure oil subspace V_{ow} of the oil-water coexistence space. The pure oil space V_o can be subdivided into the pure oil subspace V_{o1} and the pure oil subspace V_{o2}. The initial oil content of space V_{o1} is v_{o1}, the oil content of space V_{o2} is v_{o2}, and that of subspace V_{ow} is S_{or}^H. The capillary number curve has three limit values, N_{cc}, N_{ct2}, and N_{ct1}, in the case of the limit capillary number N_{cc}, only the crude oil of space V_{o1} is started and emptied, and the limit residual oil saturation is S_{or}^L. In the case of the limit capillary number N_{ct2}, the crude oil of space V_{o2} is started and emptied, and the limit residual oil saturation is S_{or}^H. For displacement when the capillary number is higher than the limit capillary number N_{ct1}, the crude oil of space V_{ow} is started.

As the capillary number curve was determined, the oil content of each subspace of the core was determined as follows:

Initial oil saturation $S_o=1-S_{wr}^L$
Initial oil content of space $V_o=v_{o1}+v_{o2}=S_o-S_{or}^H$
Initial oil content v_{o1} of space $V_{o1}=S_o-S_{or}^L$
Initial oil content v_{o2} of space $V_{o2}=S_{or}^L-S_{or}^H$
Initial oil content v_{ow} of space $V_{ow}=S_{or}^H$

Through study, we can conclude that the initial water saturation S_{wr}^L, the limit residual oil saturation S_{or}^L, S_{or}^H and the limit capillary number N_{cc}, N_{ct2}, N_{ct1} are six common key parameters of the core microscopic oil-water distribution model and the capillary number experimental curve.

5.3 Summery on the experimental method of the capillary number experimental curve

The oil displacement experiment research of the core capillary number curve is indispensable for the research and application of ASP flooding technology. The digital study of the capillary number experimental curve determines the experiment objective for the oil displacement experiment of the core capillary number curve: to determine the six key parameters of the initial water saturation S_{wr}^{L}, the limit residual oil saturation S_{or}^{L}, S_{or}^{H} and the limit capillary number N_{cc}, N_{ct2}, N_{ct1}. According to the more comprehensive understanding of the capillary number curve and the lessons from the experimental study, the following suggestions were made for the experimental determination of the capillary number curve.

In Fig. 5.1, the part that the capillary number is less than the limit capillary number N_{ct1} is the core part of the capillary number experimental curve, which must be made by the oil displacement experiment. Five basic parameters of ASP flooding can be determined from the curve. The specific experimental design will be redesigned in conjunction with the experimental curve *QL* experiment. This part of the curve can be divided into three sections: the slight decrease section of the residual oil saturation while the capillary number N_c is less than the limit capillary number N_{cc}-curve section 1; the sharp decrease section of the residual oil saturation while the capillary number N_c is between the limit capillary number N_{cc} and N_{ct2}-curve section 2; and the unchanged section of the residual oil saturation while the capillary number N_c is between the limit capillary number N_{ct2} and N_{ct1}-curve section 3.

In the capillary number experimental curve *QL* experiment, Table 5.1 shows that the whole curve was obtained from 17 oil displacement experiments completed on 17 cores. It is recommended to complete with 5 cores and about 20 experiments, including 4 experiments to determine the phase permeability curve. The design scheme is shown in Table 5.6.

Five oil displacement experiments are designed on the first core. 1-1 experiment is a water flooding experiment (refer to the 3-9 core experiment design in Table 5.1). The 1-2, 1-3 experiments are designed as polymer flooding experiments, and the 1-4, 1-5 experiments refer to 3-8 core experiment design, the flushing fluid is taken as the surfactant in the system, and the system has a relatively high interfacial tension. The 1-1, 1-2, 1-3 experiments should determine the trend of curve section 1. The 1-4, 1-5 experiments determine the trend of the upper half of curve section 2, and the position of the limit capillary number N_{cc} is basically determined at the intersection of the two lines.

Five experiments also are designed on the second core. The 2-1 experiment is to determine the water flooding (polymer flooding) phase permeability curve; the interfacial tension of the system is 7.50×10^{-2} mN/m, and the viscosity of the system is 10 mPa s. Four ASP flooding experiments of 2-2 through 2-5 are designed by changing the flooding velocity. The 2-2, 2-3 experiments should

TABLE 5.6 Experiment redesign scheme of capillary number experimental curve *QL*.

Core experiment no.	Reference experiment no.	Injecting rate (mL min^{-1})	Displacing rate (10^{-5} $m \cdot s^{-1}$)	System viscosity (mPa s)	Interfacial tension (mN/m)	V/σ_{ow} (10^{-4} $(mPa\,s)^{-1}$)	Capillary number	Residual oil saturation (%)	Corresponding experiment
1-1	3-9	1	2.919	0.6	2.25×10^{1}	0.0129	7.78×10^{-7}	28.6	
1-2		1	2.919	10	2.25×10^{1}	0.0129	1.30×10^{-5}	28.5	
1-3		2.5	7.298	10	2.25×10^{1}	0.0324	3.24×10^{-5}	28.3	
1-4	3-8	0.3	0.876	10	1.20×10^{0}	0.0730	7.30×10^{-5}	27.9	
1-5	3-8	0.5	1.460	10	1.20×10^{0}	0.1216	1.22×10^{-4}	27.4	
2-1		2.5	7.298	10	2.25×10^{1}	0.0324	3.24×10^{-5}	28.3	Relative permeability test of polymer flooding
2-2		0.5	1.460	10	7.50×10^{-2}	1.9460	1.95×10^{-3}	23.38	
2-3		0.75	3.649	10	7.50×10^{-2}	2.9190	2.92×10^{-3}	21.06	
2-4		1.10	3.211	10	7.50×10^{-2}	4.2810	4.28×10^{-3}	18.5	
2-5		1.25	3.649	10	7.50×10^{-2}	4.8650	4.87×10^{-3}	18.5	
3-1		2.5	7.298	10	2.25×10^{1}	0.0324	3.24×10^{-5}	28.3	Relative permeability test of polymer flooding

3-2		1.05	3.065	10.	7.50×10^{-2}	4.0866	4.09×10^{-3}	18.5	
4-1		0.8	2.335	15	6.50×10^{3}	35.926	5.39×10^{-2}	18.5	Relative permeability test of polymer flooding
4-2		0.85	2.481	15	6.50×10^{3}	38.172	5.73×10^{-2}	18.5	Injecting oil
4-3		0.855	2.496	15	6.50×10^{3}	38.396	5.76×10^{-2}	18.5	Injecting oil
4-4		0.860	2.510	15	6.50×10^{3}	38.621	5.79×10^{-2}	18.5	Injecting oil
4-5		0.865	2.525	15	6.50×10^{3}	38.845	5.83×10^{-2}	18.8	Injecting oil
5-1		0.850	2.481	15	6.50×10^{3}	38.172	5.73×10^{-2}	18.5	Relative permeability test of polymer flooding
5-2		0.860	2.510	15	6.50×10^{3}	38.621	5.79×10^{-2}	18.5	Injecting oil
5-3		0.8625	2.518	15	6.50×10^{3}	38.733	5.81×10^{-2}	18.7	Injecting oil
6	3-5	0.60	1.813	10.90	1.50×10^{-3}	12086	1.32×10^{-1}	25.1	
7	3-69	0.60	1.746	10.90	1.50×10^{-3}	11640	1.27×10^{-1}	25.7	
8	3-81	0.60	1.728	15.10	1.50×10^{-3}	11520	1.74×10^{-1}	20.0	
9	3-31	0.60	1.755	15.10	1.50×10^{-3}	11700	1.77×10^{-1}	20.6	
10	3-3	0.60	1.724	21.90	1.70×10^{-3}	10141	2.22×10^{-1}	23.7	
10*		**0.68**	**1.954**	21.90	1.70×10^{-3}	**11494**	$\mathbf{2.52 \times 10^{-1}}$	**23.7**	
11	3-48	0.60	1.705	21.90	1.70×10^{-3}	10029	2.20×10^{-1}	23.9	

Continued

TABLE 5.6 Experiment redesign scheme of capillary number experimental curve *QL*—cont'd

Core experiment no.	Reference experiment no.	Injecting rate (mL min^{-1})	Displacing rate (10^{-5} $m \cdot s^{-1}$)	System viscosity (mPa s)	Interfacial tension (mN/m)	V/σ_{ow} (10^{-4} $(mPa\,s)^{-1}$)	Capillary number	Residual oil saturation (%)	Corresponding experiment
11*		**0.68**	**1.932**	21.90	1.70×10^{-3}	**11365**	$\mathbf{2.49 \times 10^{-1}}$	**23.9**	
12	3-1	0.60	1.761	37.40	1.70×10^{-3}	10359	3.87×10^{-1}	19.6	
12*		**0.68**	**1.995**	37.40	1.70×10^{-3}	**11735**	$\mathbf{4.39 \times 10^{-1}}$	**19.6**	
13	3-26	0.60	1.749	37.40	1.70×10^{-3}	10288	3.85×10^{-1}	15.9	
13*		**0.68**	**1.982**	37.40	1.70×10^{-3}	**11659**	$\mathbf{4.36 \times 10^{-1}}$	**15.9**	

The bold values in core experiments 10*, 11*, 12*, and 13* are modified with interfacial tension and flooding velocity by 1.133 times.

determine the trend of the lower half of the curve section 2, the experimental point of 2-4, 2-5 experiments should fall to the curve section 3, and the position of the point (N_{ct2}, S_{or}^{H}) is determined by the intersection of the two lines.

Two experiments are designed on the third core. The 3-1 experiment is a repetition of 2-1 experiment, and 3-2 experiment is a modification of 2-4 experiment, to further confirm the position of the point (N_{ct2}, S_{or}^{H}).

The number of experiments on the fourth core is subject to the experimental data. The 4-1 experiment is to test the parameters of the ASP flooding phase permeability curve. The capillary number predesigned should be larger. The experimental point falls to the right of the midpoint on the straight section of the curve. The follow-up experiment is completed by gradually increasing the flooding velocity. Before the follow-up experiment, an appropriate amount of crude oil must be added. The experimental oil production is equal to the oil injection, indicating that there is no change in the residual oil content in the core. The experiment is repeated until the oil production is less than the oil injection, the residual oil in the core increases, and the experiment is terminated.

On the fifth core, the 5-1 experiment is a repetition of the 4-1 experiment, which is to test the parameters of the ASP flooding phase permeability curve. The subsequent experiment is to determine the limit capillary number N_{ct1} more accurately. The crude oil is injected after the 5-1 experiment, and the 4-4 experiment is repeated. Then, the experiment is repeated by increasing the flooding velocity by "small steps" until the core residual oil saturation increases, and the experiment is terminated.

It should be noted that before each oil displacement experiment, the interfacial tension and the viscosity of the system should be finally tested. If any deviation is found, the flooding velocity should be corrected, in order to ensure that the capillary number is as accurate as possible.

For the oil displacement experiment in the condition that capillary number is greater than the limit capillary number N_{ct1}, during the oil displacement experiment, the oil displacement process enters the oil-water coexistence space, and the "bound oil" originally stored in the bound oil subspace is activated and flows out, and the "flowing oil" captured in the oil-water flowing space is transformed to the "bound oil." The situation is complicated, and each experiment must be completed independently. It is recommended to design three curves using the results of the digital experiment research, to realize that as the capillary number increases, the residual oil saturation increases or decreases, and to make clear the change law of the corresponding curve distribution as the V/σ_{ow} value changes. In at least one of the three curves, the corresponding residual oil saturation value is lower than the residual oil saturation value S_{or}^{H} after the capillary number is sufficiently large.

The curve II_3 of the capillary number experimental curve QL did not synthesize "a line." In order to summarize the lessons, this group of experiments was analyzed and studied. The data in Table 5.6 were copied from Group II_3 of Table 5.1. The four experimental design schemes shared the same surfactant concentration. The interfacial tension of the system was assumed to be the same. The deviation of the interfacial tension of the system was detected before the oil displacement experiment of the last two pairs, which is a normal situation that the change of the system viscosity affects the interfacial tension. In those days, it was an exploratory experiment with no experience. Facing such situation before the experiment, the experiment can only be continued under such condition. At present, the capillary number digital study helps realize that the curve has a good coincidence with the same V/σ_{ow} value. Based on this understanding, for the eight cores in Table 5.6, we found the change of the interfacial tension of the system before the latter four core experiments, and adjusted the flooding velocity, so that the V/σ_{ow} value was close to the first four core experiments. Adjustment example: The measured interfacial tension of No. 10 experiment was 1.70×10^{-3} mN/m, which was 1.133 times of the measured interfacial tension 1.50×10^{-3} mN/m before the first two groups of core experiments. Given this, the experimental flooding velocity was expanded by 1.133 times, to calculate the V/σ_{ow} value as 1.1494 $(\text{mPa s})^{-1}$, which was very close to the experimental data of Cores 7 and 8. The experimental residual oil saturation was changed after the adjustment, and the experimental result was more ideal.

This group of experimental implementation schemes is recommended: For the displacing fluid of Group 1 experiment system, the oil displacement experiment is completed on Cores 6 and 7. The ideal experimental results should be that the corresponding displacing velocity is similar, the V/σ_{ow} value is similar, the capillary number is similar in the oil displacement process, and the residual oil saturation is similar at the end of the oil displacement process; the interfacial tension of Group 2 experimental system is basically the same as that of Group 1. The Core 8 experiment can be completed according to the original design experiment. At the end of the experiment, if the V/σ_{ow} value calculated has a relatively large deviation from the two corresponding experimental results of Group 1, the experiment should be stopped to examine the core quality. If it is similar to the result of at least one core in experiments of Group 1, the result is exactly the case that the V/σ_{ow} value of Core 8 experiment is similar to the corresponding value of Core 7 experiment. The average of the V/σ_{ow} value of Nos. 7 and 8 experiments is calculated to be 1.16 $(\text{mPa s})^{-1}$, which can be determined as the V/σ_{ow} value of the determination curve. The experimental results of the two groups can be taken as the satisfactory experimental result. The Core 6 experiment shows that the displacing velocity is relatively larger, because the pore area A value of the core interface is relatively small, and the V/σ_{ow} value is relatively large; although the residual oil saturation of the No. 6 experiment is similar to that of No. 7 experiment, the results of the Core 6 experiment are not desirable because of the large deviation in the displacing condition; the originally designed No. 9 experiment of Group 2 can be omitted, and the subsequent experiments are completed as mentioned previously. The relatively large deviation of the interfacial tension of the system is measured before No. 10 experiment, and the flooding velocity is adjusted. Scheme 10* is designed to complete the oil displacement experiment. After the experiment, the V/σ_{ow} value is calculated, and the qualified is a successful experiment, otherwise, the new experiment should be completed by repeating the above process. In this way, the oil displacement experiments can be completed on at least five cores, and four satisfactory experimental points can be obtained.

The type I capillary number experimental curve was made in the experiment, and it was not difficult to convert the type II experimental curve. The parameters obtained from the type II experimental curve were used in the digital oil displacement study.

5.4 Further perfection of the ASP flooding phase permeability curve

Section 5.1 of the literature[1] provided a formula of the ASP flooding phase permeability curve *QL* that matches with the experimental curve *QL* of the relationship between the capillary number and the residual oil, and section 5.3 gave "two formulas of the phase permeability curve." The experimental curve *QL* of the relationship between the capillary number and the residual oil, and the ASP flooding phase permeability curve *QL* have been written into the ASP flooding numerical simulation software IMCFS, and have been tested in a large number of computational studies. The core microscopic oil-water distribution model was constructed, which deepened the understanding of the irreducible water saturation S_{wr} and the residual oil saturation S_{or} in the oil displacement process, so that the oil-water phase permeability curve of the ASP flooding process can be more accurately described. First, the new basic description format was given.

Description on parameters selected for the phase permeability curve:

(1) Parameters of the water flooding (polymer flooding) phase permeability curve: the endpoint value K_o^L, K_W^L of the phase permeability curve, and the exponential value n_o^L, n_W^L of the phase permeability curve were determined by the parameter determination experiment of the water flooding phase permeability curve.

(2) Parameters of the ASP flooding phase permeability curve: the endpoint value K_o^H, K_W^H of the phase permeability curve, and the exponential value n_o^H, n_W^H of the phase permeability curve were determined by the parameter determination experiment of the ASP flooding phase permeability curve.

(3) The initial water saturation S_{wr}^L, the limit residual oil saturation S_{or}^L, S_{or}^H, and the limit capillary number N_{cc}, N_{ct2}, N_{ct1} were determined by the oil displacement experiment of the capillary number experimental curve.

(4) The displacing condition conversion parameter T_1, T_2 were obtained in the field experiment or the laboratory experiment.

The S_{or} indicated that the corresponding residual oil saturation, while the water phase saturation S_w satisfied $S_{wr}^L \leq S_w \leq 1 - S_{or}$, the water phase "normalized" saturation can be calculated by the following equation:

$$S=\frac{S_w-S_{wr}^L}{1-S_{wr}^L-S_{or}} \tag{5.8}$$

$$S_{or}=\begin{cases} S_{or}^L & N_c\le N_{cc} \\ S_{or}^L-\dfrac{N_c-N_{cc}}{N_{ct2}-N_{cc}}\left(S_{or}^L-S_{or}^H\right) & N_{cc}<N_c<N_{ct2} \\ S_{or}^H & N_{ct2}\le N_c\le N_{ct1} \\ \dfrac{S_{or}^L+T_2\cdot N_c\cdot S_{or}^H}{1+T_1\cdot N_c} & N_c>N_{ct1} \end{cases} \tag{5.9}$$

The water and oil phase permeability curve can be expressed as follows respectively:

$$K_o^0=K_w^0(S_{nw})^{n_w} \tag{5.10}$$

$$K_{or}=K_o^0(1-S_{nw})^{n_o} \tag{5.11}$$

The different variation range of the corresponding capillary number N_c, the value of the key parameters of the ASP flooding phase permeability curve:

Oil phase permeability curve

a. Endpoint value:

$$K_o^0=\begin{cases} K_o^L & N_c\le N_{cc} \\ K_o^H & N_{cc}<N_c \end{cases} \tag{5.12}$$

b. Exponential value:

$$n_o=\begin{cases} n_o^L & N_c<N_{cc} \\ n_o^H & N_{cc}<N_c \end{cases} \tag{5.13}$$

Water phase permeability curve

a. Endpoint value:

$$K_w^0=\begin{cases} K_w^L & N_c\le N_{cc} \\ K_w^L+\dfrac{N_c-N_{cc}}{N_{ct2}-N_{cc}}\left(K_w^H-K_w^L\right) & N_{cc}<N_c<N_{ct2} \\ K_w^H & N_c\ge N_{ct2} \end{cases} \tag{5.14}$$

b. Exponential value:

$$n_w=\begin{cases} n_w^L & N_c\le N_{cc} \\ n_w^L+\dfrac{N_c-N_{cc}}{N_{ct2}-N_{cc}}\left(n_w^H-n_w^L\right) & N_{cc}<N_c<N_{ct2} \\ n_w^H & N_c\ge N_{ct2} \end{cases} \tag{5.15}$$

Issues to be explained:

(1) The change of the irreducible water saturation S_{wr}

It should be noted that Eq. (5.8) is the same as the corresponding formula in the literature,[1] but there is a deeper understanding of the irreducible water saturation here.

At the beginning of the oil displacement process, the oil and water distribution in the oil layer is shown in Fig. 5.3. All water in the oil layer is concentrated in the pure water space V_w and in the pure water subspace V_{wo} of the oil-water coexistence space.

In the case of the initial water flooding, for displacement when the capillary number is less than the limit capillary number N_{cc}, the crude oil in the pore space V_{o1} with the pore radius higher than r_{oc} is started and activated, and the limit residual oil saturation is S_{wr}^L, r_{oc} is much larger than r_o; the "primary water" in the pore space with the pore radius smaller than r_o will not be driven in the case of water flooding. They are the real "irreducible water" and the total value of the initial water saturation is S_{wr}^L.

For displacement in the case of $N_{cc} \leq N_c \leq N_{ct2}$, as the capillary number increases, the pore radius of the open pore space decreases. However, while the capillary number is lower than the limit capillary number N_{ct2}, the open space is in the pure oil space V_o with the pore radius larger than r_o, as the capillary number increases, the residual oil saturation relatively decreases, but the irreducible water saturation keeps unchanged. In the case of the capillary number equal to the limit capillary number N_{ct2}, the corresponding value of the residual oil saturation S_{or} is S_{or}^H, the irreducible water saturation is still S_{wr}^L, and the residual oil and the irreducible water coexist in the space with the pore radius less than r_o.

For displacement in the case of $N_{ct2} \leq N_c < N_{ct1}$, the pore space in the pure water space V_w with smaller pore radius is opened, and the residual oil saturation remains unchanged in the oil displacement process. The "irreducible water" saturation decreases with the increase of the pore space opening in V_w, corresponding to a defined capillary number N_c. The water saturation in the space in which the pure water space V_w is activated is determined, but the amount of the activated water is difficult to determine. Given the relativity of the limit of the irreducible water saturation, the "irreducible water" in the space in the pure water space V_w is activated, and some enters into the space with the pore radius higher than r_o. The same amount of oil and water supplements the space with the pore radius greater than r_o. We can consider that the oil flowing into the space V_w is "re-driven" to the space V_o, to "retrieve" the same amount of water. The amount of water in the pore space in V_w does not change, and the water saturation in the space with the pore radius smaller than r_o is still S_{wr}^L. This is continuously determined as the "irreducible water" saturation, such treatment is the "internal adjustment" of the oil-water flowing space, which will not affect the calculation of the oil displacement effect.

For displacement in the case of the capillary number $N_c \geq N_{ct1}$, the oil-water coexistence space with the pore radius smaller than r_w is opened, and the oil in the "bound oil" subspace is released and driven away, and the water in the "irreducible water" subspace is activated and flows, some subspace releasing the "irreducible water" captures the flowing oil and transforms it into the "bound oil." By corresponding to the determined capillary number N_c, we can obtain the relative increment $\Delta = S_{or} - S_{or}^H$ of the corresponding residual oil saturation. The relative increment Δ is the relative decrement of the irreducible water saturation S_{wr}, then we have $S_{wr}^L = S_{wr}^L - \Delta$. Following the previous treatment, the flowing water with the volume "Δ" is "included" into the irreducible water S_{wr}, and the irreducible water saturation value S_{wr} is returned to S_{wr}^L.

For the displacement while the capillary number N_c is greater than the limit capillary number N_{ct1} on the grid, during the chemical flooding, the capillary number on the same grid point is changing at any time. The irreducible water saturation S_{wr} and the residual oil saturation S_{or} have the corresponding relative variation Δ. If the capillary number N_c is less than the limit capillary number N_{ct1} in the subsequent displacement, the residual oil saturation value S_{or} is determined according to the new capillary number N_c, and the irreducible water saturation is stillS_{wr}^L.

(2) Calculation of the capillary number

Since the irreducible water saturation is determined to be constant S_{wr}^L, it is not necessary to calculate the "residual water saturation" with the "oil phase capillary number," in this way, it is unnecessary to calculate the oil phase capillary number.

(3) Prevention of abnormal situations

It is noted that in the process of displacement, there are often cases where the capillary number on the grid changes from large to small, especially in the case of the slug change of the oil displacement scheme. The displacement occurs under the condition of the high capillary number N_{ct1} in the previous time step, and the grid has a low residual oil saturation S_o^0, the corresponding water saturation is S_w^0, then it is transferred to the displacement under the condition of the low capillary number N_{c2}, the residual oil saturation corresponding to the capillary number N_{c2} is S_{or}. While S_w satisfies for $S_{wr}^L \leq S_w \leq 1 - S_{or}$, we can calculate the phase permeability curve parameters K_o^0, K_w^0, n_W and n_o. Taking $S_w = 1 - S_{or}$, it is substituted into Eq. (5.8) to calculate S_{nW}, and then into Eqs. (5.10), (5.11) to calculate the phase permeability value K_{rW}, K_{ro}. The two values are directly applied to the grid for calculation corresponding to the low residual oil saturationS_o^0.

(4) Example of the phase permeability curve

Table 5.7 shows the phase permeability curve parameters against the displacement while the capillary number N_c is greater than the limit capillary number N_{cc} under the basic formula. Fig. 5.5 gives the corresponding phase permeability curve.

TABLE 5.7 The relevant parameters of the relative permeability curve with different capillary number N_c.

N_c	Δ	S_{wr}	S_{or}	$1-S_{or}$	K_w^0	n_o	K_w^0	n_w
$N_c=N_{cc}$	0	0.24	0.285	0.715	1	1.95	0.255	3.75
$N_c=0.001$	0	0.24	0.2475	0.7525	1	1.95	0.534	4.35
$N_c=0.00175$	0	0.24	0.2163	0.7838	1	1.95	0.767	4.85
$N_{ct2}\leq N_c\leq N_{ct1}$	0	0.24	0.185	0.815	1	1.87	1	5.35
$N_c=0.075$	0.055	0.24	0.24	0.76	1	1.87	1	5.35
$N_c=0.22$	−0.001	0.24	0.184	0.816	1	1.87	1	5.35
$N_c=0.50$	−0.058	0.24	0.127	0.873	1	1.87	1	5.35
$N_c=1.0$	−0.104	0.24	0.081	0.919	1	1.87	1	5.35

Note: (1) Limit capillary number $N_{cc}=0.000727$, $N_{ct2}=0.00204$, $N_{ct1}=0.05$.
(2) Under the condition of $N_c>N_{ct1}$, the residual oil saturation calculation parameter $T_1=2.5$, $T_2=0$.

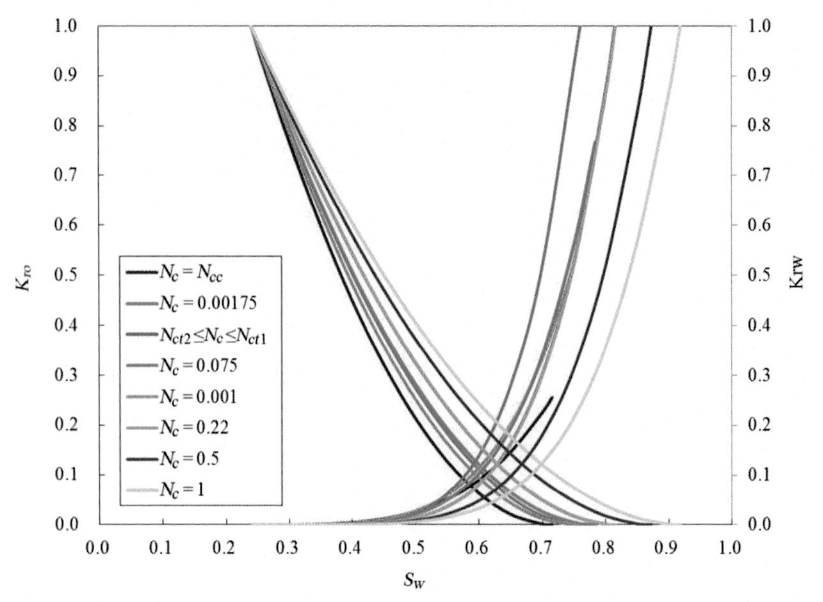

FIG. 5.5 Composite flooding relative permeability curves under different capillary number conditions.

5.5 Conclusions

(1) The experimental results of the experimental curve QL were settled through the seepage velocity formula of the classic capillary number experimental curve, to plot the type I experimental curve QL, closing the relation with the classical capillary number experimental curve. The original experimental curve QL was called the type II experimental curve QL accordingly. The seepage velocity is the water phase seepage velocity, which has important application value in digital oil displacement research.

(2) The "capillary number 'digitized' experimental curve" was obtained using the digital oil displacement experiment research method. While the capillary number N_c is less than the limit capillary number N_{cc}, there is only one capillary number curve. While the capillary number N_c is between the limit capillary numbers N_{cc} and N_{ct1}, there is a cluster of capillary number curves. Starting from point (N_{cc}, S_{or}^{L}), the curves from small to large are arranged orderly from left to right according to the V/σ_{ow} value, and at the leftmost is the "envelope line" of the curve cluster, which contains the limit capillary number N_{cc}, N_{ct2}, N_{ct1}. While the capillary number N_c is greater than the limit capillary number N_{ct1}, there is also a cluster of orderly arranged curves, whose common starting point is (N_{cc}, S_{or}^{L}).

(3) The experimental core is the objective experimental condition. The correspondence curve between the capillary number and the residual oil saturation can be obtained through a group of experiments under the determined displacement conditions. We can find the comparable displacement condition—V/σ_{ow} value, a group of experiments of the same curve have a defined V/σ_{ow} value, and the viscosity of the system increases gradually.

(4) The "classic" capillary number experimental curve is also a relation curve between the residual oil saturation and the capillary number under certain displacement condition. It corresponds to a line of the curve cluster starting from (N_{cc}, S_{or}^{L}), and has a relatively large V/σ_{ow} value. The capillary number experimental curve QL (type I) corresponds to the "envelope line" of the same cluster.

(5) 5) The capillary number experimental curve *QL* contains a group of basic parameters that completely describe the oil layer—the limit capillary number N_{cc}, N_{ct2}, N_{ct1}, the limit residual oil saturation S_{or}^{L}, S_{or}^{H}—and provides the relation on the effect of the seepage velocity, the viscosity of the system and the interfacial tension of the system on the residual oil saturation for displacement in the case greater than the limit capillary number N_{ct1}. The capillary number experimental curve QL is a relatively complete curve that is more suitable for research and application.

(6) From the core microcosmic oil-water distribution model, we summarized the experience and lessons of the experimental curve QL, and improved the experimental method of the capillary number experimental curve based on the deeper understanding of the capillary number experimental curve QL.

(7) Based on the construction of the core microcosmic oil-water distribution model, and the deep understanding of the capillary number experimental curve QL, we simplified the description of ASP flooding phase permeability curve, and the calculation in ASP flooding simulation software.

Symbol descriptions

V	displacing velocity, m/s
μ_w	displacing phase viscosity, mPa s
σ_{ow}	the interfacial tension between the displacing phase and the displaced phase, mN/m
Q	flow through the core in the unit time, cm^3/s
A	the pore area of the core section, cm^2
N_c	Capillary number
N_{cc}	the limit capillary number when the residual oil starts to flow after water flooding
N_{ct1}	the limit capillary number when the displacement condition is transformed in ASP flooding
N_{ct2}	the limit capillary number when the residual oil value corresponding to the ASP flooding process no longer decreases under the "type I" displacement
S_{or}	the residual oil saturation corresponding to the capillary number N_c

S_{wr}^{L} irreducible water saturation
S_w water phase saturation
S_o oil phase saturation
S_{or}^{L} the residual oil saturation for displacement in the case of the low capillary number, that is, the capillary number $N_c \leq N_{cc}$
S_{or}^{H} the lowest residual oil saturation during ASP flooding under the "type I" displacement, that is, the residual oil saturation corresponding to the capillary number in the case between the limit capillary number N_{ct2} and N_{ct1}
S_{nw} "Normalized" water phase saturation
T_1, T_2 displacing condition conversion parameter, which can be obtained by experiment or experimental fit
K_W^L the endpoint value of the water phase curve of the water flooding (polymer flooding) phase permeability curve
K_o^L the endpoint value of the oil phase curve of the water flooding (polymer flooding) phase permeability curve
n_W^L the exponential value of the water phase curve of the water flooding (polymer flooding) phase permeability curve
n_o^L the exponential value of the oil phase curve of the water flooding (polymer flooding) phase permeability curve
K_W^H the endpoint value of the water phase curve of the ASP flooding relative permeability curve
K_o^H the endpoint value of the oil phase curve of the ASP flooding relative permeability curve
n_W^H the exponential value of the water phase curve of the ASP flooding relative permeability curve
n_o^H the exponential value of the oil phase curve of the ASP flooding relative permeability curve
n_W the exponential value of the water phase curve of the displacement relative permeability curve under the defined capillary number N_c
n_o the exponential value of the oil phase curve of the displacement relative permeability curve under the defined capillary number N_c
T_1 displacing condition conversion parameter
T_2 displacing condition conversion parameter
V_o microscopic "pure oil" pore space with the pore radius above r_o
V_{o1} subspace of the microscopic "pure oil" pore space V_o with the pore radius above r_{oc}
V_{o2} subspace of the microscopic "pure oil" pore space V_o with the pore radius below r_{oc}
r_{oc} the pore radius at the interface of V_{o1} and V_{o2} of the microscopic "pure oil" pore space
V_w it is the "pure water" pore space when the pore radius r is between $r_o \sim r_w$

V_{ow}	it is the "pure oil" pore space when the pore radius r is in the range less than r_w
V_{wo}	it is the "pure water" pore space when the pore radius r is in the range less than r_w
r_o	the pore radius at the interface between the microscopic "pure oil" pore space V_o and the microscopic "pure water" pore space V_w
r_w	the pore radius at the interface between the microscopic "pure water" pore space V_w and the "oil-water coexistence" space
v_{o1}	initial oil content of space V_{o1}
v_{o2}	initial oil content of space V_{o2}

Reference

1. Qi LQ, Liu ZZ, Yang CZ, et al. Supplement and optimization of classical capillary number experimental curve for enhanced oil recovery by combination flooding. *Sci China Technol Sci*. 2014;57:2190–2203.

Chapter 6

Application of the microscopic oil/water distribution model in digital displacement research

Lianqing Qi[a,b], Kaoping Song[c], Hongshen Wang[b], Daowan Song[d], Jianlu Li[a], Shuai Tan[b], Shenqu Wu[b], and Ning Zhang[e]
[a]*CNPC Daqing Oilfield Exploration and Development Research Institute, Daqing, China,* [b]*CNOOC Energy Technology & Services-Drilling & Production Technology Services Co., Tianjin, China,* [c]*Northeast Petroleum University, Daqing, China,* [d]*SINOPEC Shengli Oilfield Exploration and Development Institute, Shandong, China,* [e]*CNOOC China Ltd Tianjin Branch, Tianjin, China*

Chapter outline

Development and Application of Classical Capillary Number Curve Theory
https://doi.org/10.1016/B978-0-12-821225-7.00006-0

Chapter 4 of this volume proposed a microscopic oil/water distribution model of core for oil displacement experiment and scientifically explained the complex forms of the capillary number experiment curve *QL*.[1] The crude oil in different microscopic pore spaces of the oil production cores corresponds to different capillary number ranges on the capillary number curve during displacement. This finding is used in the research on digital displacement technology to investigate the mechanisms and displacement results of different chemical displacement technologies. In this chapter, the authors discuss the main results of such investigation.

6.1 Deep understanding of reservoirs based on the innovative research findings of the capillary number experiment curve

6.1.1 Oil layer cores have the same microscopic oil/water distribution as experimental cores

Fig. 6.1 shows a capillary number experiment curve *QL* that is divided into four parts by the three capillary number limit values during displacement, namely, the critical capillary number N_{cc} during water displacement, the limit capillary number N_{ct2} during compound displacement, and the limit capillary number N_{ct1}. Fig. 6.2 shows the microscopic oil/water distribution model of experimental cores. By descending order of microscopic pore radii, the pore space of the oil layer is divided into four parts: the pure-oil space V_o, the pure-water space V_w, the oil/water coexistence space, and the pure-water or pure-oil subspace respectively. The pure-oil space V_o with a pore radius greater than r_o is divided into two pure-oil subspaces V_{o1} and V_{o2} according to the implied radius r_{oc}: V_{o1} with a pore radius greater than r_{oc}, and V_{o2} with a pore radius between r_{oc} and r_o.

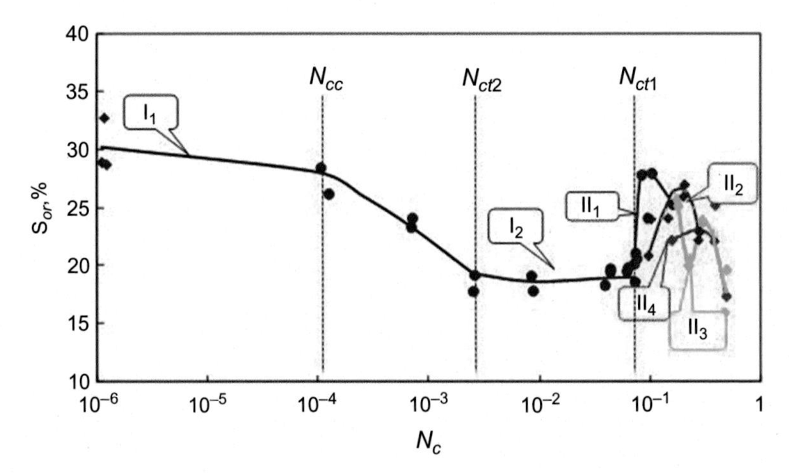

FIG. 6.1 Relation curve of residual oil saturation and capillary number (*QL*).

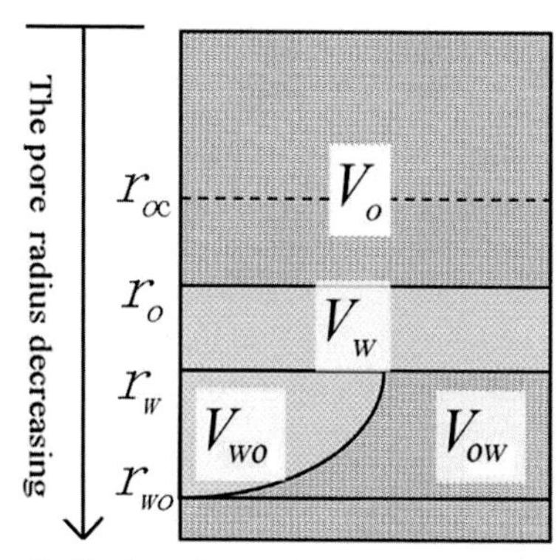

FIG. 6.2 Initial oil/water distribution scheme of experimental core.

The pure-water space V_w with a pore radius between r_o and r_w, and the oil/water-coexistence space with a pore radius between r_w and r_{wo} comprise the pure-oil subspace V_{ow} and the pure-water subspace V_{wo}. The portion with a pore radius less than r_{wo} can either be a pure-water subspace or a pure-oil one, depending on core wettability. During oil displacement, as the capillary number increases, the pore radius of the displaceable oil and/or water space decreases. The produced oil is the crude oil in the pure-oil space V_{o1} when the capillary number during water displacement is less than the critical capillary number N_{cc}. In the case of compound displacement, the produced oil is the crude oil in the pure-oil space V_{o2} when the capillary number is less than the limit capillary number N_{ct2}. The produced fluid is the water in the pure-water space V_w when the capillary number is less than the limit capillary number N_{ct1}. In the case that the capillary number is greater than the limit capillary number N_{ct1}, you may see the oil and water being displaced in the oil/water coexistence space with a pore radius less than r_w, and the pure-water subspace trapping the flowing oil.

The microscopic oil/water distribution model derived from the research on experimental cores should be applicable to the research on microscopic oil/water distribution of underground reservoir. This is because the experimental core is constructed similarly to the reservoir core so that they share nearly the same structure and physical composition and similar seepage flow property parameters. It is also because the saturation process of core oil and water is built by referring to the underground reservoir storage process. The saturated oil and water of the oil field is used in the displacement test. It's important to point out that initially the reservoir keeps only water, but later, due to changed underground pressure, the high formation pressure causes the crude oil to migrate to the reservoir where it undergoes a long "aging" process before the reservoir is developed. Similar core structure and composition and similar crude oil storage history constitute the bases to confirm that experimental core and oilfield reservoir share similar oil/water distribution.

6.1.2 Investigating oil recovery during displacement by starting with the core-based microscopic oil/water distribution

The digital geological model built by fitting the compound displacement tests described in Chapter 2 leads to the microscopic oil/water distribution model of reservoir that in turn enriches the contents of the former model. Of the displacement experiment core, the irreducible water saturation S_{wr}^{L} is measurable, so the initial oil saturation $S_{oi}=1-S_{wr}^{L}$. The residual oil saturation S_{or}^{H} with the capillary number between the limit capillary numbers N_{ct2} and N_{ct1} can also be measured, and S_{or}^{H} is the oil content v_{ow} of the oil/water space and that with smaller pore radii, and the oil content in the pure-oil pore space V_o is $v_o=1-S_{wr}^{L}-S_{or}^{H}$. For the reservoir geology model, the irreducible water saturation S_{wr}^{L} and the residual oil saturation S_{or}^{H} can be acquired by fitting field tests. The Xingerxi Block research[2] employs a 3D and three-layer simplified geology model that is able to describe the major geological characteristics of the Xingerxi Block. Table 6.1 lists the main relevant data of the digital geology model after filling in the oil contents v_o and v_{ow} of microscopic pore spaces.

On the Xingerxi Block digital geology model, we run the displacement test scheme of the Xingerxi Block. Normally, a grid's remaining oil saturation S_{or} is used to analyze the displacement result, but here the same purpose is realized by analyzing the changes of the grid's v_o and v_{ow}.

First, it is important to understand the oil distribution in different microscopic space before water flooding development. According to the data in Table 6.1, the v_o values of the three oil layers are 60.5%, 63.5%, and 65.5% respectively, and the v_{ow} values are 15.5%, 14.5%, and 13.5% respectively. The capillary number is far less than the limit one N_{cc} in the case of water displacement, and only the oil in the large-radius pores of the pure-oil space V_o can be displaced, while the flowing oil and water in these pores can be doing nothing more than flowing. The oil volume v_o in the space V_o changes during displacement. Table 6.2 lists the oil volumes v_o distributed in the grids when the compound displacement test starts. We can see that the minimum oil content is 19%, located in one of the grids of the high-permeability layer water well; the maximum oil content is 28.2%, located at the core of the low-permeability layer. As the water flooding process does not touch the oil/water coexistence space, grid oil content v_{ow} of the three-layer intervals remains 15.5%, 14.5%, and 13.5% respectively.

Tables 6.3 and 6.4 list the capillary numbers and the water phase viscosities of the planar grids of the oil layers when the main slug of the compound system is done. According to Table 6.3, the compound system solution shows clearly different spread ranges in the three layers. The capillary numbers within these spread ranges are basically between N_{ct2} and N_{ct1}, except those in the grids of the high-permeability layer water well, which are greater than N_{ct1}. It can be seen from Table 6.4 that the affected parts of the corresponding compound system have higher water phase viscosity.

TABLE 6.1 Parameters table of digital geology model of Xingerxi test area.

Oil layer	Permeability (10^{-3} μm^2)	S_{wr}^{L} (%)	S_{or}^{L} (%)	S_{or}^{H} (%)	v_o (%)	N_{cc}	N_{ct2}	N_{ct1}	T_o
Upper layer	100	24.0	36.5	15.5	60.5	0.0001	0.0025	0.0712	1.25
Middle layer	215	22.0	34.5	14.5	63.1	0.0001	0.0025	0.0712	5.0
Bottom layer	525	21.0	32.5	13.5	65.5	0.0001	0.0025	0.0712	20.0

TABLE 6.2 Distribution of remaining oil content V_o in the grids of the layers when the Xingerxi test gets started (%).

Xingerxi		1	2	3	4	5	6	7	8	9
Low-permeability layer	1	21.2	21.3	21.5	21.9	22.5	23.1	24.0	25.2	28.2
	2	21.3	21.4	21.7	22.0	22.5	23.2	24.0	25.2	28.2
	3	21.5	21.7	21.9	22.3	22.7	23.3	24.1	25.2	28.1
	4	21.9	22.0	22.3	22.6	23.0	23.5	24.2	25.2	28.1
	5	22.4	22.5	22.7	23.0	23.3	23.7	24.3	25.2	28.0
	6	23.1	23.2	23.3	23.5	23.7	24.0	24.3	25.1	27.8
	7	24.0	24.0	24.1	24.2	24.2	24.3	24.5	24.9	27.4
	8	25.1	25.1	25.2	25.1	25.1	25.1	24.9	25.0	26.5
	9	28.1	28.1	28.1	28.0	27.9	27.7	27.4	26.4	26.5
Middle-permeability layer	1	20.1	20.2	20.3	20.5	20.8	21.3	21.8	22.7	25.1
	2	20.2	20.2	20.4	20.6	20.9	21.3	21.8	22.7	25.1
	3	20.3	20.4	20.5	20.7	21.0	21.4	21.9	22.7	25.1
	4	20.5	20.6	20.7	20.9	21.2	21.5	21.9	22.7	25.0
	5	20.8	20.9	21.0	21.2	21.4	21.7	22.0	22.7	24.9
	6	21.3	21.3	21.4	21.5	21.6	21.8	22.1	22.6	24.8
	7	21.8	21.8	21.9	21.9	22.0	22.1	22.2	22.5	24.5
	8	22.7	22.7	22.7	22.7	22.7	22.6	22.5	22.6	23.8
	9	25.0	25.0	25.0	25.0	24.9	24.8	24.5	24.0	23.8
High-permeability layer	1	19.0	19.1	19.1	19.2	19.4	19.6	19.8	20.2	21.2
	2	19.1	19.1	19.2	19.3	19.4	19.6	19.8	20.2	21.2
	3	19.1	19.2	19.2	19.3	19.4	19.6	19.8	20.2	21.2
	4	19.2	19.3	19.3	19.4	19.5	19.7	19.8	20.2	21.2
	5	19.4	19.4	19.4	19.5	19.6	19.7	19.9	20.2	21.2
	6	19.6	19.6	19.6	19.7	19.7	19.8	19.9	20.1	21.1
	7	19.8	19.8	19.8	19.8	19.9	19.9	19.9	20.1	21.0
	8	20.2	20.2	20.2	20.2	20.1	20.1	20.1	20.1	20.8
	9	21.2	21.2	21.2	21.2	21.1	21.1	20.9	20.6	20.7

TABLE 6.3 Distribution of the capillary numbers in the grids of the layers when the main slug of the combination system injection of the Xingerxi test is done.

Xingerxi		1	2	3	4	5	6	7	8	9
Low-permeability layer	1	0.0315	0.0166	0.0106	0.0071	0.0029	0.0000	0.0000	0.0000	0.0000
	2	0.0166	0.0131	0.0098	0.0071	0.0012	0.0000	0.0000	0.0000	0.0000
	3	0.0106	0.0098	0.0081	0.0054	0.0007	0.0000	0.0000	0.0000	0.0000
	4	0.0071	0.0071	0.0054	0.0004	0.0000	0.0000	0.0000	0.0000	0.0000
	5	0.0029	0.0012	0.0007	0.0000	0.0000	0.0000	0.0000	0.0000	0.0000
	6	0.0000	0.0000	0.0000	0.0000	0.0000	0.0000	0.0000	0.0000	0.0000
	7	0.0000	0.0000	0.0000	0.0000	0.0000	0.0000	0.0000	0.0000	0.0000
	8	0.0000	0.0000	0.0000	0.0000	0.0000	0.0000	0.0000	0.0000	0.0000
	9	0.0000	0.0000	0.0000	0.0000	0.0000	0.0000	0.0000	0.0000	0.0000
Middle-permeability layer	1	0.0683	0.0360	0.0232	0.0167	0.0123	0.0080	0.0033	0.0000	0.0000
	2	0.0360	0.0282	0.0213	0.0163	0.0123	0.0080	0.0019	0.0000	0.0000
	3	0.0232	0.0213	0.0181	0.0149	0.0113	0.0081	0.0021	0.0000	0.0000
	4	0.0168	0.0164	0.0149	0.0124	0.0096	0.0055	0.0002	0.0000	0.0000
	5	0.0124	0.0123	0.0115	0.0096	0.0069	0.0008	0.0000	0.0000	0.0000
	6	0.0080	0.0081	0.0086	0.0054	0.0008	0.0000	0.0000	0.0000	0.0000
	7	0.0032	0.0019	0.0024	0.0002	0.0000	0.0000	0.0000	0.0000	0.0000
	8	0.0000	0.0000	0.0000	0.0000	0.0000	0.0000	0.0000	0.0000	0.0000
	9	0.0000	0.0000	0.0000	0.0000	0.0000	0.0000	0.0000	0.0000	0.0000
Middle-permeability layer	1	0.1680	0.0884	0.0568	0.0409	0.0308	0.0227	0.0147	0.0061	0.0000
	2	0.0884	0.0690	0.0519	0.0397	0.0309	0.0234	0.0162	0.0083	0.0000
	3	0.0568	0.0519	0.0439	0.0363	0.0297	0.0236	0.0171	0.0094	0.0000
	4	0.0409	0.0398	0.0364	0.0320	0.0275	0.0228	0.0173	0.0021	0.0000
	5	0.0308	0.0309	0.0297	0.0275	0.0245	0.0202	0.0158	0.0003	0.0000
	6	0.0227	0.0234	0.0235	0.0228	0.0202	0.0159	0.0014	0.0000	0.0000
	7	0.0147	0.0162	0.0170	0.0172	0.0157	0.0018	0.0000	0.0000	0.0000
	8	0.0061	0.0083	0.0094	0.0012	0.0003	0.0000	0.0000	0.0000	0.0000
	9	0.0000	0.0000	0.0000	0.0000	0.0000	0.0000	0.0000	0.0000	0.0000

TABLE 6.4 Distribution of the water phase viscosities in the grids of the layers when the injection of main slug of the combination system of the Xingerxi test is done (mPa s).

Xingerxi		1	2	3	4	5	6	7	8	9
Low-permeability layer	1	29.92	30.73	31.04	30.08	25.17	10.86	3.614	1.174	0.6827
	2	30.73	30.98	30.94	29.45	23.81	9.175	3.403	1.176	0.6824
	3	31.04	30.94	30.24	27.62	15.72	7.446	2.838	1.15	0.6829
	4	30.07	29.45	27.62	19.39	9.543	4.455	2.063	1.055	0.6797
	5	25.19	23.82	15.72	9.55	5.309	3.057	1.547	0.9642	0.6688
	6	10.88	9.192	7.449	4.46	3.059	1.853	1.238	0.8789	0.6539
	7	3.634	3.409	2.864	2.085	1.554	1.241	1.02	0.8265	0.6531
	8	1.179	1.179	1.152	1.057	0.9679	0.8825	0.8285	0.7791	0.6711
	9	0.6833	0.6832	0.6838	0.6803	0.6695	0.6535	06528	0.6716	0.6688
Middle-permeability layer	1	29.42	30.27	30.79	30.95	30.24	27.32	20.04	8.612	1.93
	2	30.27	30.57	30.87	30.87	29.9	26.73	19.54	8.253	1.888
	3	30.79	30.87	30.93	30.57	28.74	25.23	15.85	7.503	1.734
	4	30.94	30.87	30.57	29.33	26.85	22.97	13.45	4.936	1.535
	5	30.24	29.9	28.75	26.85	24.43	16.7	8.91	3.887	1.262
	6	27.33	26.75	25.31	23.03	16.62	10.76	6.264	2.958	1.019
	7	20.12	19.64	16.16	13.59	8.945	6.318	3.669	2.073	0.8929
	8	8.646	8.266	7.512	4.978	3.933	2.999	2.072	1.426	0.9007
	9	1.951	1.91	1.751	1.548	1.28	1.006	0.8884	0.8958	0.9026
High-permeability layer	1	28.89	29.66	30.27	30.66	30.83	30.49	28.72	22.26	6.715
	2	29.66	30.01	30.38	30.68	30.77	30.33	28.38	21.72	6.287
	3	30.27	30.38	30.58	30.73	30.65	29.97	27.55	20.56	4.898
	4	30.65	30.67	30.72	30.68	30.33	29.22	25.89	17.04	4.32
	5	30.81	30.76	30.63	30.31	29.54	27.39	23.29	13.51	3.219
	6	30.47	30.31	29.94	29.18	27.35	24.7	19.31	8.68	2.315
	7	28.72	28.37	27.51	25.83	23.23	19.59	12.97	5.715	1.746
	8	22.31	21.74	20.54	16.83	13.67	8.803	6.259	4.058	1.676
	9	6.869	6.474	4.963	4.432	3.291	2.455	2.052	2.091	1.725

According to Ref. 1, if the displacement is carried out in the case that the capillary number is no greater than the limit one N_{ct2}, only the crude oil in the space V_o can be displaced, and the displaced oil is flowing in a part of the opened portion of the space V_o. When the capillary number gradually increases to the limit one N_{ct1}, still only the crude oil in the space V_o can be displaced, the space where the water and oil can be flowing is the entire space V_o and a part of the opened space V_w. When the capillary number is equal to the limit one N_{ct1}, the entire space V_w is opened. In the case that the capillary number is between N_{ct2} and N_{ct1}, the opened crude oil spread range of the space V_o extends to the space V_w when the oil content v_o includes the crude oil flowing in the spaces V_o and V_w of corresponding grids. After the capillary number is greater than the limit one N_{ct1}, the oil in the pure-oil subspace V_{ow} and the water in the pure-water subspace V_{wo} are activated in the oil/water coexistence space. The activated range increases with capillary number increases, the oil and water flow into the activated part in the oil/water coexistence space, during the oil displacement process the crude oil in oil pore space is driven away by water, there is also flow oil captured by water pore space. When the oil displacement process is terminated, the remaining oil not only contains oil-bearing pore space retention and some crude oil that has not been activated, but also contains flowing oil trapped by water-bearing space that has been opened.

Based on the preceding analysis, the displacement process is terminated, and the grid's remaining oil saturation S_{or} can be further analyzed and processed. A simple method of remaining oil distribution data table in micro space of oil layer grid is established. If the capillary number is always no greater than the limit of capillary number N_{ct1}, in the case of displacement, the oil in the oil/water coexistence space V_{ow} of the grid reservoir is not activated, the oil content $v_{ow} = S_{or}^{H}$. The oil content V_o remaining in the grids' pure oil space V_o and pure water space V_w is equal to $S_{or} - S_{or}^{H}$. In this case, only the residual oil value v_o is given in the corresponding grid, and the oil content v_{ow} is omitted. When the capillary number of displacement is greater than limit capillary number N_{ct1}, if S_{or} is equal to S_{or}^{H}, then crude oil remaining in the grid could be set to zero, that is the v_o. If S_{or} is less than S_{or}^{H}, then the crude oil remaining in the grid of pure oil space V_o and pure water space V_w, that is oil content $v_o=0$, while the remaining oil in reservoir oil/water coexistence space, that is oil content $v_{ow}=S_{or}$; when the S_{or} is greater than S_{or}^{H}. The values of $S_{or} - S_{or}^{H}$ are placed in the corresponding grid, and the data are further analyzed and processed. The distribution of the grid's remaining oil content is listed in Table 6.5 when oil displacement ends. By analyzing the middle-permeability layers and according to the capillary number data given in Table 6.3, the maximum capillary number during the displacement is 0.0683, which is less than the limit capillary number N_{ct1}. Therefore only the oil in the space V_o is displaced, the oil in the oil/water coexistence space is not activated, and the flowing oil flows in the entire spaces V_o and partial space V_w. When the displacement ends, the total remaining oil content in these two pore spaces is v_o.

In Table 6.5, the content $v_o$in the water well grid is zero, indicating that the oil in these two spaces is displaced. From the grid toward the previous, the

TABLE 6.5 Distribution of the remaining oil contents in the grids of the layers when the Xingerxi test ends (%).

Xingerxi		1	2	3	4	5	6	7	8	9
Low-permeability layer	1	0.2	0.3	0.5	0.7	1.3	3.5	18.7	24.1	35.1
	2	0.3	0.4	0.5	0.8	1.4	4.2	17.3	24.4	34.8
	3	0.5	0.5	0.7	1.0	1.6	5.1	17.0	25.0	34.1
	4	0.7	0.8	1.0	1.3	2.0	7.8	18.1	25.9	34.0
	5	1.3	1.4	1.6	2.0	4.3	10.9	20.4	26.9	33.7
	6	3.7	4.0	4.9	7.7	10.7	16.5	24.8	28.2	33.4
	7	18.6	17.1	16.9	18.0	20.2	24.6	28.2	28.1	33.1
	8	24.1	24.3	24.9	25.8	26.9	28.1	28.0	28.5	31.8
	9	35.0	34.8	34.2	34.0	33.7	33.3	33.1	31.8	32.3
Middle-permeability layer	1	0	0.2	0.3	0.4	0.5	0.8	1.4	5.3	20.3
	2	0.2	0.2	0.3	0.4	0.6	0.8	1.3	3.0	15.0
	3	0.3	0.3	0.4	0.5	0.6	0.8	1.3	2.4	12.3
	4	0.4	0.4	0.5	0.6	0.7	0.9	1.4	2.3	10.9
	5	0.5	0.5	0.6	0.7	0.8	1.0	1.5	2.4	8.2
	6	0.8	0.8	0.8	0.9	1.0	1.3	1.9	2.5	7.0
	7	1.4	1.3	1.3	1.3	1.6	1.9	2.4	2.6	6.8
	8	5.3	2.9	2.4	2.3	2.4	2.6	2.6	2.8	6.1
	9	20.0	14.7	12.2	10.7	8.1	7.1	6.9	6.3	24.9
High-permeability layer	1	4.7	0.4	0.6	0.7	0.6	0.5	0.5	0.8	3.3
	2	0.4	0.5	0.7	0.7	0.5	0.4	0.5	0.7	2.4
	3	0.6	0.7	0.8	0.6	0.4	0.4	0.5	0.7	2.7
	4	0.7	0.7	0.6	0.4	0.4	0.4	0.5	0.7	2.7
	5	0.6	0.5	0.4	0.4	0.4	0.4	0.5	0.7	2.6
	6	0.5	0.4	0.4	0.4	0.4	0.4	0.5	0.7	2.4
	7	0.5	0.5	0.5	0.5	0.5	0.5	0.5	0.6	2.1
	8	0.7	0.7	0.7	0.7	0.7	0.6	0.6	0.6	1.7
	9	3.2	2.4	2.6	2.6	2.5	2.2	2.9	1.5	1.9

remaining oil content v_o slowly expands with low remaining oil values within a considerable scope; the marginal positions at one side of oil well and the oil well grids are the tail of the detained "oil wall," having high remaining oil contents. According to research, the capillary number of the middle-permeability layers is between the limit capillary numbers N_{ct2} and N_{ct1} for a long period. This "Type I" drive interval corresponding to the remaining oil saturation limit value S_{or}^{H} and indicating a good displacement result. Here, the "standard" data field of v_o value are given when the capillary number of displacement is less than the limit capillary number N_{ct1}. Now, we move on to the analysis of the low-permeability layers. For water wells, less than half of the layers receives the "Type I" drive, having lower remaining oil content; close to oil wells, more than half of the layers has high remaining oil content, and that place is the retention "oil wall" position. Special attention should be given to the high-permeability layers. From the capillary number data given in Table 6.3, when the injection of main slug of the compound system is done, a phenomenon that the grids' capillary number around the water well is greater than the limit capillary number N_{ct1} occurs. However, their values are not great and involve only a small scope. During the subsequent progress, as the slug surfactant concentration decreases, the grids' capillary number decreases. The "Type II" displacement condition that the capillary number is greater than the limit capillary number N_{ct1} only shows up within a smaller scope in a relatively short time, so all layers get a long-term "Type I" drive.

It can be seen from the calculated residual oil saturation data of the grid of the high-permeability layers that the remaining oil content S_{or} is only 4.7% for some grids around the water wells, far less than the limit remaining oil saturation S_{or}^{H} of 13.5%. The remaining oil content marked in dark green number equal to the grid remaining oil content v_{ow}, and corresponding remaining oil content v_o is zero, which is omitted. The remaining oil content $S_{or}$of other grids is higher than the limit residual oil saturation S_{or}^{H}. We calculate the values of $S_{or} - S_{or}^{H}$, then place the data in the corresponding grids. On the oil well side, it shows the values not affected by "Type II" driving conditions; the value stands for the grid remaining oil content v_o, which is marked in red. Because the values of v_o usually are smaller, it indicates that the displacement status in the grid space V_o is in good condition, and the oil in the grid V_{ow} is not driven, and the oil content v_{ow} does not change; it remains at the initial value 13.5% and it is also omitted. The changes of the numbers marked in red are that it decrease gradually from the oil well to the water well, and the minimum value is 0.4%. Reviewing the grid adjacent water well, it can be judged that the remaining oil content v_o is zero according to the remaining oil content v_o and v_{ow} of water well grid that is 0% and 4.7%, respectively. The remaining oil content in Table 6.5 is the amount of residual oil above the limit residual oil saturation S_{or}^{H}; it is the amount of flowing oil captured in space V_{ow} under "Type II" driving status and it is marked in light green. Notice that the value increases along the driving direction, and the ahead grids approach the area marked in red, so it can be judged that the residual oil content v_o in these grids is less than 0.4% and there leaved flowing oil captured by the space of V_{ow}. Further, grids marked in orange are called the "fuzzy" area of residual oil in the micro space. The data presented here of remaining oil distribution field of high-permeability layer can be used as a "standard" field data at termination with a local scope; a short term of "Type II" displacement process.

The preceding study, from the analysis of the corresponding relationship between the capillary number and reservoir microscopic space in the process of oil displacement, clearly shows the capillary number basically control near the limit number of capillary N_{ct1} in the process of displacement, the produced crude oil in the displacement process is from the pure oil reservoir microscopic pure-oil pore space, and the remaining oil distribution at different positions of the microscopic pore space at the oil displacement process termination.

6.2 Comparative studies of the displacement effects of different chemical displacement methods

By following the research and application of the chemical displacement technology designed for Daqing Oilfield, we carried out the calculations related to different chemical displacement programs applied to the updated Xingerxi Block digital geology model in order to comparatively study their displacement effects. See Table 6.6 for the main economic and technical indicators of the schemes. The schemes used the unified requirement for liquid injection speed for Daqing Oilfield; the chemical displacements with different producer-injector spacing used the same liquid injection intensity that is equal to 0.15 PV in the case of the producer-injector spacing being 250 m. The schemes realized the annual injected liquid of 0.24 PV by using the same producer-injector spacing as the Xingerxi Block. The schemes ended when the water cut of the oil wells reached 98%.

6.2.1 Research on a water displacement scheme

Water displacement can be considered a special chemical displacement method. Water displacement is an indispensable initial development stage of oilfield development through which people on field sites understand reservoirs and acquire critical technical parameters for development as well as fit and correct geological parameters for the sake of digital displacement test research. In addition, the indicators of oil recovery enhanced by chemical displacement schemes are determined according to those of oil recovery developed by water displacement schemes. Therefore studying water displacement schemes is necessarily required for performance of studying chemical displacement schemes. By calculating water displacement Scheme 1, the maximum capillary number during the oil displacement is 1.4×10^{-7}, far less than the limit capillary number $N_{cc} = 1.0 \times 10^{-4}$, and only the crude oil in the large-radius pores of the space

TABLE 6.6 Contrast of scheme characteristic parameters and oil displacement effect data.

Scheme	Scheme's liquid injection pore volume (PV)	Chemical displacement system parameter		Average of the remaining oil contents v_o in microscopic spaces (%)			Recovery factor (%)	Increment range of recovery factor (%)	Consumption of polymer equivalent to chemical agent (t)	Increment oil/ton-equivalent polymer (t/t)
		Max. viscosity (mPa s)	*Max. capillary number*	*Upper layer*	*Middle layer*	*Bottom layer*				
Water displacement 1	2.18	0.6	1.4×10^{-7}	46.00	41.23	35.78	47.21			
Water displacement 2	9.825	0.6	1.3×10^{-7}	39.83	36.68	33.45	52.80	5.59		
Polymer displacement	1.40	15.96	4.1×10^{-6}	39.63	36.48	32.95	53.63	6.42	66.52	62.33
Xingerxi block area test area	10.85	30.73	0.1682	31.92	1796	14.32	72.45	25.25	271.2	59.89
Compound displacement 1	1.79	25.37	0.1197	29.79	17.96	14.28	73.59	26.38	259.8	65.58
Polymer-compound displacement	2.48	21.42	0.0920	33.76	20.63	14.87	70.28	23.07	311.0	47.91
Water displacement 3	2.12	0.6	2.4×10^{-7}	45.75	41.37	35.82	47.24			
Compound displacement 2	2.92	32.25	0.1725	25.25	15.85	13.83	76.45	29.21	121.4	59.78

V_{o1} can be recovered during the displacement in the space V_o. When the displacement ends, a high average remaining oil content v_o remains in the three layers, and the scheme's recovery percentage is 47.21%. In order to meet the national demand for oil under the early development conditions, people wouldn't shut down a well as long as it yielded enough oil. In the test area of Xinger District, the water content of the oil wells before the compound displacement already reached greater than 99%; the pretest water displacement development stage of the central well in the test area is considered water displacement Scheme 2. According to Table 6.6, compared to Scheme 1, the amount of water injected increased by 7.65 PV when the scheme ended, the recovery percentage increased by 5.59%, and the water content of the oil wells were 99.82% when the scheme ended. Considering the conditions at that time, such a displacement result was highly acceptable. The average remaining oil content v_o of the three layers decreased respectively by 6.17%, 4.55%, and 2.33%. Table 6.2 shows the distribution data of remaining oil content v_o in the layer grids when this scheme ends. The high remaining oil contents imply a bright future for the research on enhanced oil recovery technology.

6.2.2 Research and application of polymer displacement technology

The research on enhancing oil recovery of the Daqing Oilfield began in the 1970s and reached its climax in the 1980s. According to this research, polymer displacement can significantly enhance oil recovery compared to water displacement. Polymer displacement was initially applied to the appropriate layers of all plants to the north of the First Oil Production Company. This practice is based on the research findings[3, 4] achieved at that time, namely, polymer displacement is recommended when the reservoir's coefficient of heterogeneity variation V_k is around 0.72 or greater, which brings better oil recovery than water displacement. The Xinger District had a V_k of 0.59, beyond the scope of the early polymer displacement research. The polymer displacement scheme is designed here for comparative study. As the reservoirs of Xingerxi Block are relatively homogeneous, the injection pressure rises rapidly during the polymer injection, and the safe injection pressure of the reservoirs is relatively low. Here the polymer displacement scheme uses a slug concentration of 1500 mg/L with the volume of the slug being 0.48 PV, and the safe injection pressure being the same as the compound displacement test of Xinger District. The calculations of the scheme are listed in Table 6.6. Table 6.7 gives the distribution data of the remaining oil content v_o of the three layers when the scheme ends.

According to Table 6.6, the maximum capillary number in the layers in the case of the polymer displacement scheme is 4.1×10^{-6}, which is one order of magnitude higher than that the maximum capillary number 1.4×10^{-7} in the case of the water displacement scheme and far less than the limit capillary number N_{cc}. The residual oil saturations S_{or} of the two schemes are positioned at the

TABLE 6.7 Distribution data of the remaining oil content V_o of the reservoir grids when the polymer displacement scheme ends (%).

Xingerxi		1	2	3	4	5	6	7	8	9
Low-permeability layer	1	21.1	21.1	21.2	21.4	21.6	21.9	22.6	24.6	**31.8**
	2	21.1	21.2	21.3	21.4	21.6	21.9	22.6	24.6	**31.6**
	3	21.2	21.3	21.4	21.5	21.7	22.0	22.7	24.6	**31.3**
	4	21.4	21.4	21.5	21.6	21.8	22.1	22.9	24.7	**30.9**
	5	21.6	21.6	21.7	21.8	22.0	22.3	23.1	24.8	**30.6**
	6	21.9	21.9	22.0	22.1	22.3	22.7	23.4	24.8	**30.2**
	7	22.6	22.6	22.7	22.8	23.1	23.4	23.7	24.6	**29.7**
	8	24.6	24.6	24.6	24.7	24.8	24.7	24.6	24.8	**27.8**
	9	**31.7**	**31.6**	**31.3**	**30.9**	**30.6**	**30.1**	**29.7**	**27.8**	**28.2**
Middle-permeability layer	1	20.0	20.1	20.1	20.2	20.3	20.4	20.6	21.0	23.0
	2	20.1	20.1	20.2	20.2	20.3	20.4	20.6	21.0	22.9
	3	20.1	20.2	20.2	20.3	20.3	20.4	20.6	21.0	22.9
	4	20.2	20.2	20.3	20.3	20.4	20.5	20.6	21.0	22.9
	5	20.3	20.3	20.3	20.4	20.4	20.5	20.7	21.0	22.8
	6	20.4	20.4	20.4	20.5	20.5	20.6	20.7	21.0	22.7
	7	20.6	20.6	20.6	20.6	20.7	20.7	20.8	21.0	22.6
	8	21.0	21.0	21.0	21.0	21.0	21.0	21.0	21.1	22.3
	9	23.0	22.9	22.9	22.8	22.8	22.7	22.6	22.3	22.5
High-permeability layer	1	19.0	19.0	19.1	19.1	19.1	19.2	19.3	19.4	20.4
	2	19.0	19.1	19.1	19.1	19.2	19.2	19.3	19.4	20.4
	3	19.1	19.1	19.1	19.1	19.2	19.2	19.3	19.4	20.4
	4	19.1	19.1	19.1	19.2	19.2	19.2	19.3	19.4	20.3
	5	19.1	19.2	19.2	19.2	19.2	19.3	19.3	19.4	20.3
	6	19.2	19.2	19.2	19.2	19.3	19.3	19.3	19.4	20.2
	7	19.3	19.3	19.3	19.3	19.3	19.3	19.4	19.4	20.2
	8	19.4	19.4	19.4	19.4	19.4	19.4	19.4	19.5	20.0
	9	20.4	20.4	20.4	20.3	20.2	20.2	20.1	19.9	20.0

relatively straight water displacement section of the capillary number curve, showing minor relative difference, the analysis of the capillary number determining that the polymer displacement fails to achieve a good result in enhancing oil recovery. People have concluded an applicable definition measure according to displacement practice, namely, the recovery percentage is defined as the recovery of polymer displacement when the water cut of an oil well converted to polymer solution injection well falls and then rises up to 98%; the difference well undergoing water displacement reaches 98% and the recovery of polymer displacement is defined as incremental range of recovery factor or enhanced oil recovery percentage. According to such definitions, in Table 6.6, the oil recovery factor of the water displacement scheme is 47.21% and the polymer displacement oil recovery factor of is 53.63%, which is 6.42% greater than the water displacement.

The mechanism of polymer flooding to improve oil recovery is analyzed. When the polymer displacement scheme is converted to injection of polymer solution, the water displacement has treated 0.66 PV, the water cut of oil well reaching 87.74%, the recovery percentage 37.31%. Following polymer slug, clear water is injected until the water cut of the oil well reaches 98%; the amount of injected solution in the period from starting injecting polymer solution to ending its injection is 0.75 PV, and the recovery percentage is 16.32%. In the case of water displacement Scheme 2, when the amount of injected water reaches 0.66 PV, 9.17 PV of water is continuously injected until the water cut of oil well reaches 99.98%; the recovery percentage during the injection of 9.17 PV water is 15.52%. The contrast tells us that from the same moment on, the polymer displacement scheme involves 0.75 of injected liquid and shows a recovery percentage of 16.32%, which is higher than water displacement Scheme 2's 9.17 PV and 15.52%. "Polymer displacement can recover the equal oil within a short time that water displacement needs to take a longer time to recover." Compared to Table 6.6, the separate layered residual oil value of the polymer flooding scheme and the water flooding Scheme 2, the low-, middle-, and high-permeability layers of the polymer flooding scheme are 24.13%, 20.96%, and 19.45%, respectively. These percentages are 0.2%, 1.22% and 0.5% lower than that of the water flooding Scheme 2. As shown in Table 6.7, the minimum value of the water well grids in the high-permeability layer is 19.0%, and the maximum value of the marginal grids is 20.4%, which is a difference of 1.4%. In the middle-permeability layer, the minimum value of the water well grids is 20.0%, and the maximum value of the marginal grids 23.0%, which is a difference of 3.0%. For the low-permeability layer, the "oil wall" holdup positions in the layer enriched by the high-viscosity polymer solution is marked in bold, behind which is the good displacement position by the polymer solution, where the difference between the maximum and the minimum remaining oil contents is 3.7%. As shown in Table 6.2, the minimum value of the water well grids in the high-permeability layer is 19.0%, and the maximum value of the marginal grids is 21.2%, which is a difference of 2.1%. In the middle-permeability layer, the minimum value of the water well grids is 20.1%, and the maximum value of the marginal grids is 25.1%, which is a difference of 5.0%. In the low-permeability layer, the minimum value of the remaining oil content water well grids is 21.2%, and the maximum value of the marginal grids is 28.2%, which is a difference of 7.0%. The relatively even planar distribution of remaining oil content is the result of high-viscosity polymer solution displacement expanding planar spread. During polymer displacement, the layers form "oil walls," showing the displacement result featuring changed mobility ratio and accelerated oil phase flow rate. The polymer displacement expanding planar spread effect, changing mobility ratio, and accelerating oil phase flow rate are the main mechanisms for polymer displacement enhancing oil recovery.

The successful application of the polymer displacement test for Daqing Oilfield and other oilfields in the 1980s is an important achievement of chemical displacement technology for enhanced oil recovery. The industrial application of the polymer displacement technology at Daqing Oilfield was extraordinarily significant for China when the country needed to use crude oil in exchange for foreign currencies. The slogan of the research on enhanced oil recovery of Daqing Oilfield at that time was that to increase 1% of crude oil production in the oil recovery of Daqing Oilfield is equivalent to discovering another Yumen Oilfield. The polymer displacement test successfully addressed the pressing needs of China. By carrying out large-scale industrial production based on polymer displacement, Daqing Oilfield made a significant contribution to the country.

6.2.3 Research and application of compound displacement technology

Almost at the same time as researching and applying polymer displacement technology, Daqing Oilfield carried out research on compound displacement technology of ternary system of ASP. After successful laboratory research, field tests were conducted, from small pilot test to expanded test, and then industrial tests and applications were carried out. The successful field tests prove that composite flooding is an effective, new technology to greatly improve oil recovery. Daqing Oilfield carried out the research and applications of ASP displacement technology while carrying out those of the polymer displacement technology. The field tests successfully proved that the compound displacement as a high-tech technology can significantly enhance oil recovery. The compound displacement test performed in Xinger District of Daqing Oilfield is a relatively early pilot-scale test. Table 6.6 shows the main results of the scheme, namely, that compared to polymer displacement, water displacement resulted in an increase of oil recovery by 25.25%. Compared to polymer displacement, water displacement resulted in an increase of oil recovery by 18.73%. The average remaining oil content of all the layers respectively decreased by 7.71%, 18.50%, and 18.63%. According to the distribution data of the remaining oil content of all the layers at the end of the test given in Table 6.5, compound displacement achieves a good result. The compound system solution having entered the low-permeability layer is held up in its middle portion when the scheme ends, in front of which is a high-oil-bearing "oil wall," which shows that the held-up compound system slug should be pushed forward continuously in order to further enhance the oil recovery. According to this, relevant calculations were made to obtain an improved scheme: the compound system slug is shrunk to 0.3 PV followed by polymer slug of 0.45 PV. Table 6.6 lists the results of the improved version of compound displacement Scheme 1. The contrast shows that its displacement result is better than the field scheme and its oil recovery factor increases by 1.23% relatively and to 26.18% compared to water displacement that is nearly 20% higher than the polymer displacement scheme. The amount of chemical agent used is equivalent to 11.4 tons of less used polymer, and the amount of oil increased by the ton-equivalent polymer is 5.69 tons compared to the field test; the average remaining oil content v_o of the three layers decreases respectively by 9.84%, 18.50%, and 18.67% compared to the polymer displacement layers. Table 6.8 lists the distribution data of the remaining oil content v_o when compound displacement Scheme 1 ends. Tables 6.9–6.11 list the capillary numbers of the grids of the three layers of all the reservoirs at different times. A study on them is given in what follows.

First of all, let's analyze the data of the high-permeability layer listed in Table 6.9. When the injection of the compound system finished, the compound system solution affected more than half of the layers; the capillary number in

TABLE 6.8 Distribution data of the remaining oil content of the reservoir grids when combination displacement Scheme 1 ends (%).

	Xingerxi	1	2	3	4	5	6	7	8	9
Low-permeability layer	1	0.3	0.5	0.7	0.9	1.2	2.6	**14.4**	**23.8**	**34.8**
	2	0.5	0.6	0.8	0.9	1.2	2.7	**10.7**	**23.9**	**33.2**
	3	0.7	0.8	0.8	1.0	1.4	2.7	9.8	**24.2**	**32.6**
	4	0.9	0.9	1.0	1.2	1.8	4.0	**10.2**	**25.4**	**32.5**
	5	1.2	1.2	1.4	1.8	2.5	5.3	**14.9**	**27.0**	**32.3**
	6	2.6	2.6	2.7	3.8	5.4	**10.8**	**21.4**	**27.1**	**31.9**
	7	**15.0**	**10.4**	9.3	9.9	**14.6**	**21.0**	**26.0**	**27.9**	**31.8**
	8	**23.8**	**23.8**	**23.9**	**25.4**	**27.3**	**27.0**	**27.5**	**28.2**	**31.1**
	9	**35.0**	**33.6**	**32.7**	**32.5**	**32.0**	**31.6**	**31.8**	**31.1**	**31.5**
Middle-permeability layer	1	0.2	0.3	0.5	0.7	3.8	0.9	1.3	3.8	**17.7**
	2	0.3	0.4	0.5	0.7	2.5	0.9	1.3	2.5	**11.7**
	3	0.5	0.5	0.6	0.7	2.2	0.9	1.2	2.2	8.6
	4	0.6	0.7	0.7	0.8	2.2	0.9	1.2	2.2	7.4
	5	0.8	0.8	0.8	0.8	2.2	1.0	1.2	2.2	7.9
	6	0.9	0.9	0.9	0.9	2.5	1.1	1.5	2.5	7.8
	7	1.3	1.3	1.2	1.2	2.6	1.5	1.9	2.6	7.4
	8	3.7	2.4	2.1	2.2	2.8	2.5	2.6	2.8	6.1
	9	**17.6**	**11.4**	8.2	7.4	6.3	7.7	7.4	6.3	**24.2**
High-permeability layer	1	9.7	0.2	0.2	0.3	0.4	0.6	0.6	0.8	2.5
	2	0.2	0.2	0.3	0.4	0.5	0.6	0.6	0.7	2.3
	3	0.2	0.3	0.3	0.4	0.5	0.6	0.6	0.7	2.4
	4	0.3	0.4	0.4	0.5	0.5	0.6	0.6	0.7	2.4
	5	0.5	0.5	0.5	0.5	0.6	0.6	0.6	0.7	2.4
	6	0.6	0.6	0.6	0.6	0.6	0.5	0.5	0.7	2.4
	7	0.6	0.6	0.6	0.6	0.6	0.5	0.5	0.6	2.2
	8	0.8	0.7	0.7	0.7	0.7	0.6	0.5	0.5	1.7
	9	2.4	2.2	2.3	2.3	2.3	2.2	2.0	1.4	1.9

TABLE 6.9 Distribution data of the capillary number in the grids of the high-permeability layers at different times during the performance of combination displacement Scheme 1.

	Xingerxi	1	2	3	4	5	6	7	8	9
When the injection of combination system finishes	1	0.1197	0.0629	0.0404	0.0291	0.0217	0.0156	0.0089	0.0007	0.0000
	2	0.0629	0.0492	0.0369	0.0282	0.0216	0.0156	0.0090	0.0009	0.0000
	3	0.0404	0.0369	0.0311	0.0256	0.0204	0.0149	0.0091	0.0003	0.0000
	4	0.0291	0.0282	0.0256	0.0222	0.0184	0.0134	0.0074	0.0000	0.0000
	5	0.0217	0.0216	0.0204	0.0185	0.0157	0.0123	0.0023	0.0000	0.0000
	6	0.0156	0.0157	0.0150	0.0135	0.0123	0.0095	0.0002	0.0000	0.0000
	7	0.0090	0.0091	0.0091	0.0077	0.0028	0.0002	0.0000	0.0000	0.0000
	8	0.0008	0.0010	0.0004	0.0000	0.0000	0.0000	0.0000	0.0000	0.0000
	9	0.0000	0.0000	0.0000	0.0000	0.0000	0.0000	0.0000	0.0000	0.0000
When the injection of polymer slug finishes	1	0.0000	0.0000	0.0000	0.0000	0.0000	0.0001	0.0023	0.0062	0.0032
	2	0.0000	0.0000	0.0000	0.0000	0.0000	0.0001	0.0034	0.0099	0.0066
	3	0.0000	0.0000	0.0000	0.0000	0.0000	0.0005	0.0052	0.0137	0.0099
	4	0.0000	0.0000	0.0000	0.0000	0.0001	0.0011	0.0093	0.0173	0.0133
	5	0.0000	0.0000	0.0000	0.0001	0.0006	0.0028	0.0173	0.0210	0.0172
	6	0.0001	0.0001	0.0005	0.0011	0.0027	0.0083	0.0237	0.0255	0.0225
	7	0.0024	0.0033	0.0051	0.0087	0.0158	0.0238	0.0271	0.0319	0.0309
	8	0.0064	0.0099	0.0136	0.0172	0.0212	0.0262	0.0322	0.0414	0.0486
	9	0.0032	0.0066	0.0098	0.0133	0.0180	0.0254	0.0381	0.0714	0.0714
When the scheme ends	1	0.0000	0.0000	0.0000	0.0000	0.0000	0.0000	0.0000	0.0000	0.0022
	2	0.0000	0.0000	0.0000	0.0000	0.0000	0.0000	0.0000	0.0000	0.0044
	3	0.0000	0.0000	0.0000	0.0000	0.0000	0.0000	0.0000	0.0000	0.0065
	4	0.0000	0.0000	0.0000	0.0000	0.0000	0.0000	0.0000	0.0001	0.0056
	5	0.0000	0.0000	0.0000	0.0000	0.0000	0.0000	0.0000	0.0002	0.0054
	6	0.0000	0.0000	0.0000	0.0000	0.0000	0.0000	0.0000	0.0003	0.0060
	7	0.0000	0.0000	0.0000	0.0000	0.0000	0.0000	0.0000	0.0004	0.0068
	8	0.0000	0.0000	0.0000	0.0001	0.0002	0.0002	0.0003	0.0008	0.0057
	9	0.0022	0.0044	0.0064	0.0057	0.0054	0.0055	0.0045	0.0029	0.0029

most grids of the affected positions is between 0.0025 and 0.0712, and only the capillary number of the water well grids is 0.1183, which indicates that the layer gets the good "Type I" drive during the displacement, corresponding to the remaining oil saturation limit S_{or}^{H}. When the injection of the subsequent polymer

TABLE 6.10 Distribution data of the capillary number in the grids of the middle-permeability layers at different times during the performance of combination displacement Scheme 1.

Xingerxi		1	2	3	4	5	6	7	8	9
When the injection of combination system finishes	1	0.0486	0.0255	0.0162	0.0111	0.0067	0.0021	0.0000	0.0000	0.0000
	2	0.0255	0.0199	0.0148	0.0108	0.0069	0.0033	0.0000	0.0000	0.0000
	3	0.0162	0.0148	0.0122	0.0092	0.0061	0.0034	0.0000	0.0000	0.0000
	4	0.0111	0.0107	0.0092	0.0069	0.0039	0.0005	0.0000	0.0000	0.0000
	5	0.0067	0.0066	0.0062	0.0039	0.0012	0.0000	0.0000	0.0000	0.0000
	6	0.0028	0.0022	0.0033	0.0005	0.0000	0.0000	0.0000	0.0000	0.0000
	7	0.0000	0.0000	0.0000	0.0000	0.0000	0.0000	0.0000	0.0000	0.0000
	8	0.0000	0.0000	0.0000	0.0000	0.0000	0.0000	0.0000	0.0000	0.0000
	9	0.0000	0.0000	0.0000	0.0000	0.0000	0.0000	0.0000	0.0000	0.0000
When the injection of polymer slug finishes	1	0.0000	0.0000	0.0000	0.0000	0.0002	0.0057	0.0051	0.0020	0.0003
	2	0.0000	0.0000	0.0000	0.0000	0.0004	0.0081	0.0057	0.0032	0.0004
	3	0.0000	0.0000	0.0000	0.0000	0.0012	0.0088	0.0065	0.0044	0.0005
	4	0.0000	0.0000	0.0000	0.0005	0.0071	0.0086	0.0070	0.0055	0.0004
	5	0.0002	0.0004	0.0011	0.0070	0.0091	0.0082	0.0074	0.0064	0.0005
	6	0.0057	0.0082	0.0090	0.0086	0.0081	0.0080	0.0078	0.0074	0.0007
	7	0.0056	0.0060	0.0067	0.0070	0.0073	0.0077	0.0082	0.0084	0.0018
	8	0.0031	0.0036	0.0045	0.0053	0.0061	0.0073	0.0083	0.0095	0.0023
	9	0.0005	0.0010	0.0016	0.0021	0.0024	0.0013	0.0010	0.0044	0.0001
When the scheme ends	1	0.0000	0.0000	0.0000	0.0000	0.0000	0.0000	0.0000	0.0008	0.0009
	2	0.0000	0.0000	0.0000	0.0000	0.0000	0.0000	0.0001	0.0013	0.0019
	3	0.0000	0.0000	0.0000	0.0000	0.0000	0.0000	0.0002	0.0018	0.0025
	4	0.0000	0.0000	0.0000	0.0000	0.0000	0.0000	0.0003	0.0019	0.0026
	5	0.0000	0.0000	0.0000	0.0000	0.0000	0.0000	0.0005	0.0025	0.0026
	6	0.0000	0.0000	0.0000	0.0000	0.0000	0.0001	0.0012	0.0034	0.0035
	7	0.0000	0.0000	0.0001	0.0002	0.0004	0.0011	0.0023	0.0037	0.0051
	8	0.0009	0.0013	0.0017	0.0018	0.0024	0.0034	0.0037	0.0020	0.0055
	9	0.0009	0.0019	0.0025	0.0026	0.0026	0.0036	0.0048	0.0057	0.0001

TABLE 6.11 Distribution data of the capillary number in the grids of the low-permeability layers at different times during the performance of combination displacement Scheme 1.

Xingerxi		1	2	3	4	5	6	7	8	9
When the injection of combination system finishes	1	0.0222	0.0116	0.0071	0.0035	0.0007	0.0000	0.0000	0.0000	0.0000
	2	0.0116	0.0091	0.0065	0.0024	0.0004	0.0000	0.0000	0.0000	0.0000
	3	0.0071	0.0065	0.0049	0.0005	0.0000	0.0000	0.0000	0.0000	0.0000
	4	0.0035	0.0027	0.0030	0.0001	0.0000	0.0000	0.0000	0.0000	0.0000
	5	0.0007	0.0004	0.0000	0.0000	0.0000	0.0000	0.0000	0.0000	0.0000
	6	0.0000	0.0000	0.0000	0.0000	0.0000	0.0000	0.0000	0.0000	0.0000
	7	0.0000	0.0000	0.0000	0.0000	0.0000	0.0000	0.0000	0.0000	0.0000
	8	0.0000	0.0000	0.0000	0.0000	0.0000	0.0000	0.0000	0.0000	0.0000
	9	0.0000	0.0000	0.0000	0.0000	0.0000	0.0000	0.0000	0.0000	0.0000
When the injection of polymer slug finishes	1	0.0000	0.0000	0.0000	0.0004	0.0049	0.0032	0.0004	0.0000	0.0000
	2	0.0000	0.0000	0.0000	0.0011	0.0053	0.0039	0.0027	0.0000	0.0000
	3	0.0000	0.0000	0.0003	0.0063	0.0049	0.0029	0.0009	0.0000	0.0000
	4	0.0004	0.0011	0.0063	0.0053	0.0045	0.0021	0.0007	0.0000	0.0000
	5	0.0050	0.0053	0.0049	0.0044	0.0038	0.0029	0.0002	0.0000	0.0000
	6	0.0034	0.0039	0.0030	0.0020	0.0029	0.0002	0.0001	0.0000	0.0000
	7	0.0012	0.0028	0.0008	0.0006	0.0002	0.0001	0.0000	0.0000	0.0000
	8	0.0000	0.0000	0.0000	0.0000	0.0000	0.0000	0.0000	0.0000	0.0000
	9	0.0000	0.0000	0.0000	0.0000	0.0000	0.0000	0.0000	0.0000	0.0000
When the scheme ends	1	0.0000	0.0000	0.0000	0.0000	0.0001	0.0009	0.0006	0.0001	0.0000
	2	0.0000	0.0000	0.0000	0.0000	0.0003	0.0010	0.0009	0.0003	0.0000
	3	0.0000	0.0000	0.0000	0.0000	0.0007	0.0011	0.0010	0.0004	0.0000
	4	0.0000	0.0000	0.0000	0.0004	0.0012	0.0013	0.0011	0.0004	0.0000
	5	0.0001	0.0003	0.0007	0.0012	0.0014	0.0015	0.0010	0.0003	0.0000
	6	0.0009	0.0011	0.0012	0.0013	0.0015	0.0015	0.0010	0.0003	0.0000
	7	0.0006	0.0010	0.0010	0.0011	0.0011	0.0009	0.0005	0.0002	0.0000
	8	0.0001	0.0003	0.0003	0.0003	0.0002	0.0001	0.0001	0.0001	0.0000
	9	0.0000	0.0000	0.0000	0.0000	0.0000	0.0000	0.0000	0.0000	0.0000

slug finishes, the slug pushes forward to half the layers near the oil well and its front edge advances to the oil well; the capillary number of the main body position grids of the slug remains between 0.0025 and 0.0712. According to relevant analysis, the layer plane gets the best drive during the entire displacement.

According to Chapter 4, the displaced under such a drive is the crude oil in the space V_o. According to Table 6.8, when the scheme ends, the residual oil content v_o staying in pore spaces V_o and V_w of the layer grids is very low; the marginal positions at the oil well side have lower remaining oil content than the test scheme. Note that the water well grids get a long-term displacement with an extra-high capillary number in addition to the high-viscosity system, so the displacement result is excellent. The remaining oil content on the grids is marked with green digits, indicating that the remaining oil content v_o on the grids is "0" and that the numbers are the value of the remaining oil content v_{ow}. Let's analyze the middle-permeability layer given in Table 6.10. When the injection of the compound system finished, the contrast shows that the positions affected by the compound system are obviously less than the high-permeability layer. However, the capillary number in most of the grids of these affected positions is between 0.0025 and 0.0712; when the injection of the polymer slug finishes, the compound system is pushed forward to involve a considerable area, and their grids' capillary number mostly is between 0.0025 and 0.0712. When the scheme ends, the main body of the slug reaches the oil well, but its most of the oil stays in the layer. According to Table 6.8, the displacement result of the entire layer is not as good as the high-permeability layer, but is also better; the remaining oil content v_o of most grids is low except the marginal positions on the oil well side. When the injection of the compound system finishes, compared to the test scheme, the grids with a high remaining oil content decrease, as does the remaining oil content. Let's analyze the data of the low-permeability layer shown in Table 6.11. When the injection of the compound system finishes, there are 14 grids between 0.0025 and 0.0712, which is three grids less than the field test. It should be noted that the main slug volume of the field scheme is 0.35 PV while the compound system slug volume of the improved scheme is 0.3 PV. When the injection of the polymer slug finishes, the slug pushes forward to involve larger area. Nineteen grids at the main positions have a capillary number between 0.0025 and 0.0712. The place where the main body of the slug passes through shows a good displacement result. When the scheme ends, the main body of the slug enters half the layer on the oil well side; the data show that the capillary number of the grids is less than the limit capillary number $N_{ct2} = 0.0025$, indicating that the slugs are already dispersed and that the displacement result is not good. The data shown in Table 6.8 indicate that the distribution of the remaining oil content values of the grids is basically as same as the test scheme. The displacement result relatively improved, and the "oil wall" with a high oil-bearing content held up at the front positions shrinks in scope, and the remaining oil content values relatively decrease.

From the analysis, one can see the optimized compound displacement scheme can keep the capillary number between N_{ct2} and N_{ct1} within a relatively larger scope of the layers and a relatively longer time during the displacement, and recover more crude oil from the microscopic space V_o and achieve a higher oil recovery rate.

6.2.4 Knowledge of the limitations in application of the polymer displacement technology

Compound displacement can enhance oil recovery by more than 20% compared to water displacement. Relatively speaking, polymer displacement is better kept to take up around 10%. It is important to consider what to do after polymer displacement, and the answer is probably still compound displacement. In other words, the development process of an oilfield would be water displacement-polymer displacement-compound displacement. The polymer displacement before that is not only a vain effort but also may influence the compound displacement that follows it. In that case, we can decide that the performance of two displacements wouldn't lead to a result better than implementation of the compound displacement alone. Let's give the polymer-compound displacement scheme an analog calculation. After the aforementioned polymer displacement scheme is carried out, compound displacement Scheme 1 follows. Tables 6.6, 6.12, and 6.13 list the calculation results. The contrastive analysis against compound displacement Scheme 1 shows that the polymer-compound displacement scheme achieves a final recovery percentage of 70.59%, down by 3.30%; the recovery percentage of the low-, middle-, and high-permeability layers respectively decreases by 3.97%, 2.68%, and 0.58%. An extra 51.9 tons of polymer is consumed, and the ton-equivalent polymer increases 48.55 tons of recovered oil, which is less than compound displacement Scheme 1 and the polymer displacement scheme. This proves that the displacement effect of the polymer-compound displacement scheme is comparatively poorer.

TABLE 6.12 Water phase viscosity data of the layers grids when the polymer-combination displacement scheme is switched to the combination system displacement scheme (mPa s).

Xingerxi		1	2	3	4	5	6	7	8	9
Low-permeability layer	1	1.35	3.43	8.88	13.88	15.07	15.21	14.02	8.10	3.09
	2	3.43	7.57	11.66	14.34	14.98	15.03	13.82	8.19	3.32
	3	8.87	11.63	14.13	14.64	14.79	14.66	13.3	8.18	3.63
	4	13.79	14.32	14.63	14.57	14.44	14.13	12.34	8.10	3.93
	5	15.03	14.94	14.76	14.42	13.97	13.07	11.12	8.04	4.21
	6	15.2	15.00	14.63	14.09	13.01	11.56	9.93	8.02	4.52
	7	14.04	13.84	13.3	12.33	11.08	9.89	8.98	8.07	4.86
	8	8.14	8.23	8.21	8.14	8.07	8.04	8.07	7.85	6.65
	9	3.10	3.34	3.65	3.96	4.24	4.55	4.89	6.45	6.38
Middle-permeability layer	1	0.65	0.75	1.13	2.75	5.07	7.74	10.43	13.6	10.25
	2	0.75	0.86	1.33	3.25	4.90	6.18	9.52	13.34	10.42
	3	1.118	1.33	2.49	4.16	4.03	4.82	8.77	12.74	10.36
	4	2.686	3.21	4.15	3.82	3.53	4.42	8.08	11.78	10.21
	5	5.006	4.85	3.98	3.50	3.55	4.23	7.33	10.51	10.00
	6	7.688	5.95	4.72	4.30	4.14	4.36	6.02	8.95	9.69
	7	10.25	9.39	8.63	7.87	6.78	5.58	6.25	7.98	9.12
	8	13.53	13.24	12.56	11.49	10.02	8.56	7.68	7.79	8.36
	9	10.32	10.47	10.41	10.23	9.97	9.57	8.74	8.33	8.17
High-permeability layer	1	0.60	0.60	0.61	0.65	0.81	1.41	4.05	9.61	14.15
	2	0.60	0.60	0.61	0.66	0.82	1.44	3.82	9.21	13.81
	3	0.61	0.61	0.63	0.70	0.89	1.49	3.49	8.57	13.20
	4	0.65	0.66	0.70	0.76	1.05	1.54	3.17	7.87	12.43
	5	0.81	0.81	0.88	1.04	1.23	1.61	2.86	6.83	11.51
	6	1.37	1.41	1.46	1.50	1.57	1.83	2.66	4.77	10.32
	7	3.94	3.70	3.35	3.01	2.69	2.54	2.83	3.94	8.52
	8	9.47	9.04	8.35	7.58	5.84	4.36	3.69	4.01	7.26
	9	14.12	13.75	13.08	12.23	11.13	9.56	7.71	5.39	6.30

TABLE 6.13 Distribution data of the remaining oil content of the layer grids when the polymer-combination displacement scheme ends (%).

Xingerxi		1	2	3	4	5	6	7	8	9
Low-permeability layer	1	0.5	0.7	1.1	1.6	2.6	13.7	23.0	24.7	29.5
	2	0.7	0.9	1.2	1.6	2.9	16.1	23.0	24.9	29.5
	3	1.2	1.2	1.3	1.7	4.8	20.5	23.4	25.1	29.4
	4	1.6	1.6	1.7	2.6	9.0	23.1	24.4	25.5	29.3
	5	2.6	2.8	4.6	9.4	19.2	23.6	24.8	25.7	29.1
	6	13.4	15.4	20.3	23.1	23.6	24.8	25.4	25.9	28.7
	7	23.0	23.0	23.4	24.3	24.8	25.4	25.8	26.1	28.2
	8	24.7	24.8	25.0	25.5	25.7	26.0	26.2	26.3	27.8
	9	29.4	29.5	29.5	29.4	29.2	28.6	28.5	27.9	28.3
Middle-permeability layer	1	0.3	0.5	0.8	1.3	2.2	2.7	3	14.2	24.7
	2	0.5	0.6	0.9	1.4	2.4	2.9	2.9	8.3	24.6
	3	0.8	0.9	1.1	1.6	2.8	2.7	2.7	4.9	24.2
	4	1.3	1.4	1.6	2.2	3.4	2.5	2.4	4.5	22.6
	5	2.1	2.3	2.8	3.4	3.1	2.3	2.2	3.7	18.2
	6	2.8	2.9	2.8	2.5	2.3	2.2	2.3	3.5	13.7
	7	3.0	2.8	2.7	2.4	2.2	2.3	2.5	3.7	9.0
	8	13.7	7.8	4.8	4.4	3.5	3.4	3.6	4.1	7.2
	9	24.7	24.5	24.2	22.3	18.2	13.8	8.6	7.1	25.9
High-permeability layer	1	11.6	0.2	0.3	0.5	0.8	1.2	1.5	1.7	3.7
	2	0.2	0.3	0.4	0.6	0.8	1.2	1.4	1.6	3.1
	3	0.3	0.4	0.5	0.7	1.0	1.1	1.3	1.6	3.3
	4	0.5	0.6	0.7	0.9	1.2	1.1	1.2	1.5	3.3
	5	0.8	0.9	1.0	1.1	1.2	1.1	1.0	1.4	3.3
	6	1.2	1.2	1.1	1.1	1.0	0.9	0.8	1.4	3.2
	7	1.5	1.4	1.3	1.1	1.0	0.8	0.7	0.9	3.0
	8	1.8	1.6	1.6	1.5	1.4	1.3	0.8	0.7	2.2
	9	3.6	3.0	3.2	3.2	3.1	3.0	2.6	1.8	2.4

By analyzing the water phase viscosity data of the reservoir grids given in Table 6.12, when the former displacement scheme is switched to the compound system solution, we can see that the water phase of most reservoir grids has a high viscosity due to the existence of entrapped polymer solution. In the high-permeability layer, the high-viscous polymer solution has advanced to the oil well with some residues remaining at both wings of the mainstream line, some grids at the marginal positions at one side of the oil well showing a high viscosity. In the middle-permeability layer the entrapped high-viscous solution involves a larger area with much more residues remaining at both wings of the main stream line. In the low-permeability layer, a large area of high-viscous polymer solution is entrapped in the central position of the layer. Due to the retention of the polymer solution, the layer pressure rises rapidly during the injection of the combination system. In order to ensure injection safety, the viscosity of the injected system has to be lowered, which leads to a relative decrease in the volume of the compound system solution entering the low-permeability layer. As the slugs are pushed forward, the flow resistance at both sides of the main stream line of the middle- and high-permeability layers is relatively low, the displacement channels relatively clear, while the low-permeability layer involves a large resistance so that the displacement is seriously hindered plus the relatively darting of the high-permeability layer, the compound displacement consequently ends earlier, and the scheme's liquid injection multiple relatively shortens by 0.0563 PV. By analyzing the distribution data of the remaining oil content v_o of the layers given in Table 6.13, when the scheme ended, compared to the data of compound displacement Scheme 1 given in Table 6.8, the grids' remaining oil content values of the high-permeability layer are slightly higher. Those of the middle-permeability layer are obviously higher as well, especially at the marginal positions at both wings on one side of the oil well. In the low-permeability layer, the high-oil-bearing scope covers up more than half of the layers. The analysis clearly shows the reasons why the polymer-compound displacement scheme has a bad displacement result.

In order to protect resources, polymer displacement technology must be used cautiously in the future. However, the technical achievement acquired during the application of the polymer displacement to Daqing Oilfield is a priceless treasure. The high-viscous system displacement changes the mobility ratio, speeds up the oil recovery rate, and improves the planar and vertical influence effect. Daqing Oilfield applied this to the compound displacement technology test so that the polymer concentration of the Xingerxi Block Area system at the early stage hit 2300 mg/L, ensuring that the test achieves a high enhanced oil recovery result. The optimized displacement scheme recommended by Ref. 3, the compound system slugs of 0.3 PV and the subsequent polymer slug of 0.6 PV are also the results of the polymer displacement technology summarization. Such a scheme can also be referred to as "compound system slug-polymer solution slug compound displacement" scheme.

6.2.5 In-depth understanding of small-spacing high-viscous ultra-low interfacial tension system displacement result

The research in Chapter 2 optimizes the compound displacement conditions, recommending producer-injector spacing of 125 m, annual liquid injection speed of 0.4 PV, a two-stage slug, slug volume of 0.3 PV, surfactant concentration of 0.3%, system interfacial tension of 1.25×10^{-3} mN/m, and follow-up polymer slug volume of 0.6 PV; the two-stage slug to use the same polymer concentration, and the concentration value to take a highest possible value according to layer injection situation. The calculation results are listed under compound displacement Scheme 2 in Tables 6.6, 6.14, and 6.15. According to Table 6.6, the average remaining oil content v_o of the three layers is respectively 25.25%, 15.85%, and 13.83%, the scheme's recovery percentage is 76.45%, which is 29.21% higher than the water displacement's 47.24% under the same well spacing condition, and the ton-equivalent polymer increasing oil by 59.78 tons.

Table 6.14 lists the capillary number data of the reservoir grids when the injection of compound system ends. Each layer of a reservoir has 81 grids. The number of the grids with a capillary number between 0.0025 and 0.0712 is 46 in the high-permeability layer, 30 in the middle-permeability layer, and 17 in the low-permeability layer. It's also noted that there are three grids with a capillary number greater than 0.0712 near the water well in the high-permeability layer. Table 6.15 lists the data of the remaining oil content v_o of thc three layers when the program ends. The contrast of the two tables

TABLE 6.14 Capillary number data of the reservoir grids when the injection of combination system switched from combination displacement Scheme 2 ends.

Xingerxi		1	2	3	4	5	6	7	8	9
Low-permeability layer	1	0.1669	0.0883	0.0546	0.0308	0.0043	0.0000	0.0000	0.0000	0.0000
	2	0.0883	0.0670	0.0476	0.0267	0.0005	0.0000	0.0000	0.0000	0.0000
	3	0.0546	0.0476	0.0338	0.0115	0.0000	0.0000	0.0000	0.0000	0.0000
	4	0.0308	0.0267	0.0115	0.0001	0.0000	0.0000	0.0000	0.0000	0.0000
	5	0.0043	0.0005	0.0000	0.0000	0.0000	0.0000	0.0000	0.0000	0.0000
	6	0.0000	0.0000	0.0000	0.0000	0.0000	0.0000	0.0000	0.0000	0.0000
	7	0.0000	0.0000	0.0000	0.0000	0.0000	0.0000	0.0000	0.0000	0.0000
	8	0.0000	0.0000	0.0000	0.0000	0.0000	0.0000	0.0000	0.0000	0.0000
	9	0.0000	0.0000	0.0000	0.0000	0.0000	0.0000	0.0000	0.0000	0.0000
Middle-permeability layer	1	0.3507	0.1847	0.1198	0.0832	0.0521	0.0211	0.0001	0.0000	0.0000
	2	0.1848	0.1428	0.1083	0.0806	0.0519	0.0219	0.0000	0.0000	0.0000
	3	0.1199	0.1084	0.0898	0.0686	0.0458	0.0152	0.0000	0.0000	0.0000
	4	0.0832	0.0807	0.0686	0.0529	0.0321	0.0005	0.0000	0.0000	0.0000
	5	0.0522	0.0520	0.0459	0.0322	0.0016	0.0000	0.0000	0.0000	0.0000
	6	0.0212	0.0220	0.0154	0.0006	0.0000	0.0000	0.0000	0.0000	0.0000
	7	0.0001	0.0001	0.0000	0.0000	0.0000	0.0000	0.0000	0.0000	0.0000
	8	0.0000	0.0000	0.0000	0.0000	0.0000	0.0000	0.0000	0.0000	0.0000
	9	0.0000	0.0000	0.0000	0.0000	0.0000	0.0000	0.0000	0.0000	0.0000
High-permeability layer	1	0.8620	0.4506	0.2869	0.2048	0.1523	0.1077	0.0597	0.0017	0.0000
	2	0.4506	0.3506	0.2604	0.1970	0.1499	0.1074	0.0628	0.0016	0.0000
	3	0.2870	0.2605	0.2176	0.1768	0.1395	0.1010	0.0596	0.0005	0.0000
	4	0.2049	0.1971	0.1769	0.1514	0.1228	0.0875	0.0492	0.0001	0.0000
	5	0.1524	0.1501	0.1396	0.1229	0.0993	0.0727	0.0048	0.0000	0.0000
	6	0.1078	0.1076	0.1014	0.0878	0.0725	0.0335	0.0002	0.0000	0.0000
	7	0.0599	0.0630	0.0604	0.0500	0.0056	0.0002	0.0000	0.0000	0.0000
	8	0.0016	0.0016	0.0006	0.0001	0.0000	0.0000	0.0000	0.0000	0.0000
	9	0.0000	0.0000	0.0000	0.0000	0.0000	0.0000	0.0000	0.0000	0.0000

TABLE 6.15 Distribution data of the remaining oil content of the layers grids when combination displacement Scheme 2 ends (%).

Xingerxi		1	2	3	4	5	6	7	8	9
Low-permeability layer	1	0.2	0.3	0.4	0.6	0.7	1.0	3.2	**19.8**	**28.4**
	2	0.3	0.4	0.5	0.6	0.7	1.0	2.4	**15.1**	**28.9**
	3	0.4	0.5	0.5	0.6	0.7	1.0	2.0	**14.2**	**29.3**
	4	0.6	0.6	0.6	0.6	0.8	1.1	2.1	**15.1**	**29.3**
	5	0.7	0.7	0.7	0.8	1.0	1.3	2.2	**16.3**	**29.2**
	6	1.0	1.0	1.0	1.1	1.3	1.6	2.8	**17.0**	**28.6**
	7	3.3	2.4	2.0	2.1	2.2	2.8	6.9	**18.1**	**29.5**
	8	**20.1**	**15.4**	**13.9**	**14.7**	**16.2**	**17.1**	**17.9**	**19.7**	**27.8**
	9	**28.5**	**29.0**	**29.3**	**29.3**	**29.1**	**28.5**	**29.8**	**28.1**	**27.7**
Middle-permeability layer	1	0.1	0.2	0.3	0.4	0.5	0.5	0.7	1.1	8.8
	2	0.2	0.2	0.3	0.4	0.5	0.5	0.7	1.1	3.7
	3	0.3	0.3	0.4	0.4	0.5	0.5	0.7	1.1	3.9
	4	0.4	0.4	0.4	0.4	0.5	0.5	0.7	1.0	3.9
	5	0.5	0.5	0.5	0.5	0.5	0.5	0.7	1.0	3.8
	6	0.5	0.5	0.5	0.5	0.5	0.6	0.7	1.0	3.6
	7	0.7	0.7	0.7	0.6	0.7	0.7	0.7	1.1	3.4
	8	1.1	1.1	1.0	1.0	1.0	1.0	1.1	1.3	3.1
	9	8.7	3.7	3.9	3.9	3.8	3.6	3.4	3.2	**22.8**
High-permeability layer	1	7.4	11.7	0.2	0.2	0.3	0.3	0.4	0.5	1.1
	2	11.7	0.1	0.2	0.2	0.3	0.3	0.4	0.5	1.0
	3	0.2	0.2	0.2	0.3	0.3	0.3	0.4	0.4	1.0
	4	0.2	0.2	0.3	0.3	0.3	0.4	0.4	0.4	1.1
	5	0.3	0.3	0.3	0.3	0.3	0.4	0.4	0.4	1.1
	6	0.3	0.3	0.3	0.4	0.4	0.4	0.3	0.4	1.1
	7	0.4	0.4	0.4	0.4	0.4	0.3	0.3	0.3	1.0
	8	0.5	0.5	0.4	0.4	0.4	0.4	0.3	0.3	0.8
	9	1.1	1.0	1.0	1.0	1.1	1.0	0.9	0.6	0.9

can be clearly seen; in the section of high-permeability layer, the remaining oil amount v_{ow} in the grid near the well is marked in green color, displacement under the condition of high viscosity and high capillary number has been experienced there, and the oil amount in the microscopic oil/water coexistence space has been reduced. The whole front layer is well driven, and the maximum remaining oil amount v_o is only 1.1%. In the middle-permeability layer, the entire layer gets the good Type I drive, the remaining oil v_o of the marginal and corner grids being 8.8% at most, and the oil well grids being the remainder of the "oil wall" tail. In the low-permeability layer a large area gets the good "Type I" drive condition, 60% of the area being low residual oil content, and a part of the "oil wall" is entrapped in the layer.

Research confirms that in the case of small-well-spacing high-viscosity ultra-low interfacial tension (1.25×10^{-3} mN/m) system compound displacement, the capillary number during the main displacement is between the limit capillary numbers N_{ct2} and N_{ct1}; the reservoirs are under the good Type I drive, corresponding to the remaining oil saturation limit S_{or}^{H}; and the recovered crude oil basically comes from the microscopic pure-oil space V_o. Under the oil reservoir conditions of Daqing Oilfield, this displacement achieves an oil recovery rate that is 30% higher than that of water displacement.

6.2.6 Research on the digitization of ASP displacement in the Warden Unit of Sho-Vel-Tum Oilfield

Ref. 5 introduced an ASP displacement field test that had been performed on the Warden Unit of Sho-Vel-Tum Oilfield in Oklahoma, United States, in February 1998. The test employed a formula with a high concentration of surfactant and a high alkali concentration: 2.2 wt% Na_2CO_3 + 0.5 wt% ORS-62 + 1500 mg/L Alcoflood 1275A, the system's underground interfacial tension 2×10^{-5} mN/m, one well of the well test group for injection and four wells for production, producer-injector spacing of 71 m, and oil incremental range of recovery factor by only 16.22%

From the analysis of Literature 5, one can see, during the test experiment, 15 wells outside the test well group achieved synchronous effect with the test well group, which was due to the significant increase of reservoir pressure caused by the leakage of injected flooding system to the Well Group. This shows obvious characteristic of "Type II" driving condition. Further, from the data we find that there is a significant difference between the wells in Well Group 4 and injection wells, which led to significant differences of the benefits in oil wells, indicating that the test is carried out under the condition of a very serious horizontal heterogeneity reservoir, and the displacement fluid break through into the direction of the high-permeability layers and the "Type II" driver condition aggravates the thrust on the plane. In addition, the depth of the reservoirs in the test area is about 213.6 m, and for such a shallow reservoir, the high-viscosity system could not be injected into the reservoir, the test condition is not ideal. What is to be studied here is the displacement effect of ultra-low interfacial tension system, so in-depth research is carried out on the digital oil displacement geological platform of Xingerxi Block Area compound displacement obtained in the previous study instead of the original geological model of test conditions.

Now, we calculate the two displacement schemes as the injection production well spacing is 88 m.

Scheme 1: compound flooding system slug is 0.3 PV, surfactant concentration is 0.5 wt%, interfacial tension is 2×10^{-5} mN/m, polymer concentration is 1200 mg/L, working viscosity of 12 mPa s, subsequent polymer slug is 0.2 PV, polymer concentration is 1200 mg/L. (Note: it is determined that the polymer concentration of two-stage slug is changed to 1200 mg/L by fitting the flooding effect. The internal reason is the difference of polymer viscosity curve and other factors.)
Scheme 2: the polymer concentration of the former scheme compound system slug and the subsequent polymer slug is increased to 3000 mg/L, other parameters remain unchanged, and the maximum underground working viscosity of the scheme system is about 50 mPa s.

Parameters of the two schemes and oil displacement effect are listed in Table 6.16. For comparison, the test data of Xingerxi Block are given, and Table 6.17 shows the distribution of grid capillary number, system viscosity distribution, and the distribution data of residual oil amount v_o and v_{ow} at the termination time when converted to inject the compound system by high-permeability model for 228 days.

The injection rate is 1.2 PV per year. Plan Scheme No. 1 is designed with reference to the compound flooding test in Oklahoma, United States. The plan increased oil recovery by 18.84%, because the oil layer is changed to homogeneous, the flooding effect is better than the field test results. It should also explain that the scheme is in high capillary number and high residual oil saturation conditions, the results of the "fitting" to obtain a set of transformation parameters T_1, T_2 of drive status in oil layers. It not only related to the reservoir properties, but is also associated with displacement condition. Take the "correction" set of parameters for geologic model to ensure the oil displacement results under high capillary number has a relatively high credibility and accuracy.

The viscosity of Plan Scheme No. 2 is more than two times higher than that of Plan Scheme No.1, and the increase of oil recovery is 8.95% higher, which clearly shows that the high viscosity ultra-low interfacial tension system can greatly improve the recovery rate. It is noted in particular that there is a certain decrease in the residual oil content in the high- and middle-permeability oil layers. In the previous study, the goal of compound flooding is to increase oil recovery in the low permeability layer. Here, the goal of compound flooding to improve oil recovery is extended to reduce the remaining oil saturation in the high-permeability and middle-permeability layers. Based on the significant difference of oil recovery effect between the two schemes in the high-permeability layer, only the driving conditions of the two schemes are analyzed and studied here. According to Table 6.17, when the solution of the two schemes was converted to the compound slug for 228 days, both the compound system slug injected and the polymer slug injected was also finished. As shown in the table,

TABLE 6.16 Contrast of scheme characteristic parameters and oil displacement effect data.

Scheme	Implementation time (days)	Parameter of combinational system			Stratified residual oil (%)			Recovery factor (%)	Increment range of recovery factor (%)	Consumption of polymer equivalent to chemical agent (t)	Increment oil/ton-equivalent polymer (t/t)
		Interfacial tension (10^{-3} mN/m)	*Max. viscosity (mPa s)*	*Max. capillary number*	*Upper layer*	*Middle layer*	*Bottom layer*				
Experiment	1567	1.25	31.04	0.169	31.92	17.96	14.32	72.45	25.25	271.2	60.13
1	725	0.02	13.6	2.43	36.79	26.55	15.62	66.13	18.84	**77.22**	**30.31**
2	701	0.02	51.06	15.18	13.90	17.84	10.89	75.08	27.79	**93.23**	**37.04**
3	674	0.02	52.12	15.25	31.33	15.73	9.65	75.66	28.37	**67.26**	**52.35**

TABLE 6.17 Distribution data of the capillary number, system viscosity and remaining oil content of the high permeability layers grids.

item	grid	1	2	3	4	5	6	7	8	9
Capillary number of low viscosity system scheme after injecting 228 day	1	0.0000	0.0000	0.0000	0.0000	0.0000	0.0012	0.0887	0.1729	0.0866
	2	0.0000	0.0000	0.0000	0.0000	0.0000	0.0015	0.1192	0.1797	0.1710
	3	0.0000	0.0000	0.0000	0.0000	0.0001	0.0027	0.1588	0.2662	0.2749
	4	0.0000	0.0000	0.0000	0.0000	0.0004	0.0083	0.2267	0.3555	0.3956
	5	0.0000	0.0000	0.0001	0.0004	0.0024	0.0494	0.4883	0.5088	0.5341
	6	0.0011	0.0014	0.0024	0.0071	0.0405	0.5402	0.6439	0.7016	0.7185
	7	0.3607	0.1222	0.1639	0.2348	0.5104	0.6527	0.7850	0.9395	1.0018
	8	0.1215	0.1815	0.2691	0.3781	0.5425	0.7361	0.9495	1.2370	1.5460
	9	0.0895	0.1709	0.2657	0.3867	0.5405	0.7785	1.2171	2.2422	2.2422
Capillary number of high viscosity system scheme after injecting 228 days	1	0.0000	0.0000	0.0000	0.0000	0.0002	0.0067	0.4394	0.6082	0.4338
	2	0.0000	0.0000	0.0000	0.0000	0.0003	0.0120	0.5744	0.9310	0.9299
	3	0.0000	0.0000	0.0000	0.0000	0.0008	0.0544	0.8456	1.4390	1.5499
	4	0.0000	0.0000	0.0000	0.0003	0.0058	0.8953	1.5169	2.0174	2.2573
	5	0.0002	0.0003	0.0007	0.0056	0.0215	1.1786	2.2922	2.6752	3.1060
	6	0.0063	0.0111	0.0480	0.9096	1.1876	2.2527	3.0023	3.5027	4.1122
	7	0.4413	0.5831	0.8514	1.4130	2.2876	3.0373	3.7044	4.7196	5.5736
	8	0.6148	1.1504	1.4665	2.1050	2.7583	3.5797	4.6970	6.1893	8.2203
	9	0.4558	0.9626	1.6126	2.3814	3.3678	4.6468	6.8300	11.714	11.714
Subsurface system viscosity of low viscosity system scheme after injecting 228 days/mPa·s	1	0.62	0.64	0.73	1.02	1.72	3.47	3.16	8.43	7.93
	2	0.64	0.66	0.78	1.10	1.78	3.50	6.16	8.39	7.79
	3	0.73	0.78	0.97	1.37	2.16	3.74	6.15	8.34	7.67
	4	1.02	1.11	1.37	2.03	3.04	4.26	5.60	8.43	7.64
	5	1.72	1.80	2.17	3.03	3.94	4.89	7.45	8.48	7.64
	6	3.48	3.52	3.76	4.25	4.85	6.33	7.87	8.50	7.68
	7	6.18	6.18	6.16	6.73	7.39	7.82	8.31	8.54	7.73
	8	8.49	8.47	8.44	8.46	8.46	8.46	8.53	8.43	7.80
	9	8.01	7.88	7.82	7.76	7.77	7.83	7.89	7.89	7.79
Subsurface system viscosity of high viscosity system scheme after injecting 228 days/mPa·s	1	0.67	0.76	1.07	1.90	3.79	8.90	16.90	26.83	31.30
	2	0.76	0.87	1.21	2.16	3.95	9.23	16.83	26.91	30.96
	3	1.07	1.21	1.69	2.88	4.94	10.32	17.14	26.36	30.41
	4	1.88	2.15	2.87	4.22	6.78	11.77	18.59	26.85	29.99
	5	3.76	3.92	4.88	6.74	10.81	15.21	21.21	27.83	29.61
	6	8.69	9.07	10.10	11.62	18.04	18.95	24.24	28.81	29.34
	7	16.69	16.58	16.83	18.13	20.81	23.94	27.18	29.86	29.10
	8	26.26	26.12	25.96	26.34	27.19	28.26	29.37	30.33	28.83
	9	31.33	30.94	30.31	29.84	29.43	29.14	28.84	28.52	28.36
Remaining oil at the termination of the low viscosity system scheme %	1	8.91	10.06	11.13	11.66	11.68	0.13	3.00	6.63	13.29
	2	10.06	11.10	11.11	11.36	11.70	13.37	2.40	5.11	14.34
	3	11.12	11.11	11.13	11.03	11.86	13.17	1.49	4.81	12.64
	4	11.65	11.35	11.07	11.58	12.34	13.09	0.73	4.75	12.23
	5	11.66	11.70	11.88	12.46	12.65	12.99	0.23	4.80	11.05
	6	0.12	13.39	13.20	13.24	13.03	12.95	13.23	4.25	7.03
	7	2.99	2.37	1.49	0.75	0.19	13.24	12.93	2.20	7.83
	8	6.33	4.94	4.86	4.99	4.81	4.34	1.02	13.50	13.31
	9	13.18	13.45	13.00	11.53	11.32	6.41	5.89	10.89	12.92
Remaining oil at the termination of the high viscosity system scheme %	1	6.97	7.25	8.13	7.34	7.96	8.52	9.28	4.74	7.11
	2	7.25	7.95	7.39	7.37	7.89	8.38	9.06	4.20	6.32
	3	8.13	7.39	7.27	7.51	7.96	8.33	9.04	1.51	6.47
	4	7.34	7.37	7.50	7.79	8.01	8.31	8.94	0.55	6.70
	5	7.96	7.89	7.96	8.02	8.04	8.20	8.66	12.48	6.54
	6	8.51	8.37	8.31	8.29	8.15	8.03	8.22	10.85	6.20
	7	9.27	9.04	9.04	8.95	8.57	8.11	7.86	9.85	5.81
	8	4.71	3.01	1.04	13.10	11.65	10.50	9.00	9.27	5.51
	9	7.03	6.53	6.58	6.69	6.48	6.08	5.51	12.39	2.13

at that moment, two schemes have a large capillary number on the grid that is higher than the limit capillary number N_{ct1}. It shows that the two schemes are both at the "Type II" driving status. The location tagged with "red number" marker of the scheme of low-viscosity system is advanced, the number in the corresponding position is obviously small, clearly showing the low-viscosity system "Type II" drive condition in the water phase breakthrough. Meanwhile in the high-viscosity system, water phase breakthrough is restrained in "Type II" drive condition. The underground working viscosity distribution in the grids when analyzing converted compound system for 228 days, at this time, is between 5 and 10 mPa s, which is less than 57% of all the grids for the low-viscosity system scheme. For the high-viscosity system, that is 65% between 10 and 30 mPa s. As can be seen from the experimental curve *QL* of capillary number N_c, the remaining oil content drops below the limit residual oil saturation S_{or}^H only when the capillary number N_c is sufficiently greater than the limit N_{ct1} and the system has a high viscosity. Obviously, the scheme of the high-viscosity system is more suitable for such conditions. At the end of the scheme, the remaining oil in the green mark grid falls below the limit residual oil saturation S_{or}^H, which is the remaining oil remaining in the oil/water coexistence space. It can be seen from the table that there are 38 green number grids in low-viscosity system schemes. On these grids, the schemes of the high-viscosity system are also marked with green numbers and have relatively low values. In addition to the first 38 grids of the high-viscosity system scheme, another 20 grids are marked in green. Obviously, the low-viscosity system scheme of these 20 grids has relatively high residual oil saturation. In addition to the 23 grids in the above 58 grids, the remaining oil values of the two schemes are marked. By comparison, it can be seen that the high-viscosity system scheme has relatively small residual oil values. Upon analysis, the high-viscosity system scheme and the high-permeability layer have relatively better displacement effect.

Comparing the scheme of Xingerxi Block and the scheme of high-viscosity and extra-ultra-low system, the latter improved oil recovery of 2.63%, and lowered residual oil content of 2.82% in the low-permeability layer, 0.15% in the middle-permeability layer, and 3.43% in the high-permeability layer. Taking the price of surfactant as 2.5 times that of polymer, the equivalent incremental oil production per ton polymer reduces to 37.04 tons for the high-viscosity ultra-low scheme. Furthermore, if the surfactant price drops to 1.5 times than the polymer price, the equivalent incremental oil production per ton polymer increases to 52.24 tons. One question is worth discussing. The ASP flooding test of Warden Unit in the Sho-Vel-Tum Field in Oklahoma is a promotion project of the United States Department of Energy. Why it choose 0.5 wt% ORS-62, however, the previous compound flooding tests in the United States all adopted a surfactant concentration of 0.3 wt%. There may be two reasons. The first is that the surfactant ORS-62 has a large adsorption quantity in the underground oil reservoir, and in order to ensure the ultra-low underground interfacial tension achieved 2×10^{-5} mN/m, so a high concentration of surfactant was selected. Another possibility is that the designer of the scheme payed attention to the idea of "high concentration surface-active agent system has an extended sweep effect"; neither of these designs is desirable. The test using high adsorption quantity surfactant exists the higher risk, so it must not be used in the high-cost test. The relatively high viscosity of the compound system slug and subsequent polymer slug is an effective measure to expand the underground sweep effect. Based on this understanding, Scheme 3 is designed on the basis of Scheme 2, in which the surfactant concentration is reduced to 0.3 wt%, and the maximum average reservoir pressure in the flooding process is basically the same as the previous two schemes, so the polymer concentration in the slug is determined accordingly.

According to Table 6.14, the fluid maximum viscosity slightly increased when compared to Scheme 2 in the process of displacement, oil residual content of low-permeability layer is relatively increased by 1.93%, decreased by 2.08% and 1.24% for the middle-permeability layer and high-permeability layer, respectively. The amplitude of enhanced oil recovery of Scheme 3 increased to 0.58%, surfactant usage amount reduced 10.65 tons, and polymer increased by 4.95 tons. The equivalent incremental oil production per ton polymer increased by 52.35 tons; 15.31 tons compared to Scheme 2.

It is found that the high-viscosity ultra-low interfacial tension system has displaced the oil in the microscopic oil/water space, which is more difficult and has improved the effect.

6.3 Conclusions

(1) Research confirms that the microscopic oil/water distribution model of experimental core is applicable to reservoir core.
(2) Starting with the core-based microscopic oil/water distribution model, the analytical study of the digital displacement test result clearly shows the different ranges corresponding to the capillary number curve as well as the crude oil in the different microscopic pore spaces of recovered layers during the displacement.
(3) Compound flooding is an efficient oil displacement technology with enhanced oil recovery. When the high-viscosity ultra-low interfacial tension system is used, the produced oil is mainly from the microscopic pure oil space V_o, originally accumulated in the oil reservoir, which has a high recovery rate and should be widely applied. In the application process, further optimization measures should be taken so as to obtain the best oil displacement effect.
(4) Research confirms that if the high-viscosity extra-ultra-low interfacial tension system is used, the oil displacement range is extended to the microscopic oil/water coexistence space in the reservoir, and the recovery factor is improved to some extent. However, it is more difficult and costly.
(5) The produced oil of polymer flooding is only from the subspace V_{o1} of reservoir microscopic space V_o which pore radius is bigger than the others microscopic space. The oil recovery is significantly lower than compound flooding. In addition, the crude oil in the subspace V_{o2} with smaller pore radius in the space V_o could not be driven out and remains at reservoirs. Hence, polymer displacement technology should be used cautiously in the future in order to protect precious underground resources.

Symbol descriptions

μ_w	displacing phase viscosity, mPa s
σ_{ow}	interfacial tension between displaced phase and displaced phase, mN/m
"Type I" driving status	displacement under the condition that the capillary number is less than or equal to the limit capillary number N_{ct1}
N_c	capillary number
N_{cc}	the limit capillary number when the residual oil starts to flow after water flooding
N_{ct1}	the limit capillary number when the displacement condition is transformed in ASP flooding
N_{ct2}	the limit capillary number when the residual oil value corresponding to the ASP flooding process no longer decreases under the "type I" displacement
PV	the pore volume of oil layer
r_{oc}	the pore radius at the interface of V_{o1} and V_{o2} of the microscopic "pure-oil" pore space
S_o	oil phase saturation
S_{oi}	initial oil saturation
S_{or}	the residual oil saturation corresponding to the capillary number N_c
S_{or}^H	the lowest residual oil saturation during ASP flooding under the "type I" displacement, that is, the residual oil saturation corresponding to the capillary number in the case of the limit capillary number between N_{ct2} and N_{ct1}
S_{or}^L	the residual oil saturation for displacement in the case of low capillary number, that is, capillary number $N_c \leq N_c$
S_{wr}^L	bound water saturation
S_w	water phase saturation
T_1, T_2	displacing condition conversion parameter, which can be obtained by experiment or experimental fit
"Type II" driving status	displacement under the condition that the capillary number is greater than the limit capillary number N_{ct1}
V_k	reservoir heterogeneity variation coefficient
v_o	pore volume percentage of initial oil content in microscopic pore space V_o, after reservoir development, v_o includes oil volume retained in space V_o and space V_w
V_o, r_o	the pore space with a micro pore radius greater than r_o is initially "pure-oil" pore space
V_{o1}	subspace of the microscopic "pure-oil" pore space V_o with pore radius greater than r_{oc}

V_{o2}	subspace of the microscopic "pure-oil" pore space V_o with pore radius less than r_{oc}
V_{ow}, V_{wo}	V_{ow} is the "pure-oil" pore space when the pore radius r is in the range less than r_w; V_{wo} is the "pure-water" pore space when the porc radius r is in the range less than r_w
v_{ow}	pore volume percentage of initial oil content in microscopic pore space V_{ow}, after reservoir development, v_{ow} includes oil content retained in space V_{ow} and space V_{wo}
V_w, r_w	it is the "pure-water" pore space when the pore radius r is between $r_o \sim r_w$

References

1. Qi LQ, Liu ZZ, Yang CZ, et al. Supplement and optimization of classical number experimental curve for enhanced oil recovery by combination flooding. *Sci China Technol Sci.* 2014;57:2190–2203.
2. Wang D, Cheng J, Junzheng W, et al. *Summary of ASP Pilots in Daqing Oil Field SPE 57288; 1999.*
3. Wang Z, Jingcun Z, Yanli J, et al. *Evaluation of Polymer Flooding in Daqing Oil Field and Analysis of Its Favourable Conditions SPE 17848; 1988* .
4. Lianqing Q. *Numerical Simulation Research of Polymer Displacement Engineering*. Beijing: Petroleum Industry Press; 1988.
5. Felber BJ. *Selected U.S. Department of Energy's EOR Technology Applications. SPE84904; 2003:* 1–11.

Chapter 7

Further study on the digital field experiment of combination flooding

Lianqing Qi[a,b], Yi Wu[c], Kaoping Song[d], Yiqiang Li[e], Yong Shi[f], Dali Weng[b], Hongqing Zhu[b], and Jun Wei[b]

[a]*CNPC Daqing Oilfield Exploration and Development Research Institute, Daqing, China,* [b]*CNOOC Energy Technology & Services-Drilling & Production Technology Services Co., Tianjin, China,* [c]*CNPC, Liaohe Oilfield Exploration and Development Institute, Liaoning, China,* [d]*Northeast Petroleum University, Daqing, China,* [e]*Enhanced Oil Recovery Institute of China University of Petroleum, Beijing, China,* [f]*CNOOC Energy Technology & Services-Safety & Environmental Protection Co., Tianjin, China*

Chapter outline

Development and Application of Classical Capillary Number Curve Theory
https://doi.org/10.1016/B978-0-12-821225-7.00007-2

The proposal of capillary number experiment curves by American scholars in the 1950s opened up a new era of chemical displacement technology research. In the 1980s, after 30 years of concentrated study, researchers began to field test this technology with great success. Following the American scholars, petroleum researchers in China became actively involved in the studies of chemical displacement technology. They firstly experienced success in polymer displacement field experiments in the late 1980s. From the 1990s, pilot ASP displacement field experiments were successful, as were subsequent expansive experiments. Following this success, the technology began to be used in industrial applications.

7.1 A breakthrough result in theoretical studies

7.1.1 Supplementation and perfection of classic capillary number experiments

Moore 1, Taber 2, and Foster 3, among others, proposed the concept of hydrodynamic force and capillary force ratio in order to research and describe the "relationship between the hydrodynamic force of captured residual oil and the capillary force" during the displacement process. This is called the capillary number and its equation is as follows:

$$N_c = \frac{V \cdot \mu_w}{\sigma_{ow}} \tag{7.1}$$

wherein N_c stands for capillary number (without dimension), V for displacing phase seepage velocity (m/s), μ_w for displacing phase viscosity (mPa s), and σ_{ow} for interfacial tension (mN/m) between displacing phase and displaced phase. The corresponding relation curve between capillary number and residual oil is given through further experiments, commonly called "capillary number curve." Scholars obtained curves with different shapes from different angles. Fig. 7.1 shows an experimental curve finished by Moore and Slobod.

This important research is considered to be the theoretical foundation of compound displacement technology.

Based on the work of the American scholars, the authors of this volume conducted their own experiment,[4] which resulted in the discovery of "capillary number experiment curve *QL*," as shown in Fig. 7.2. Comparing Figs. 7.1 and 7.2,

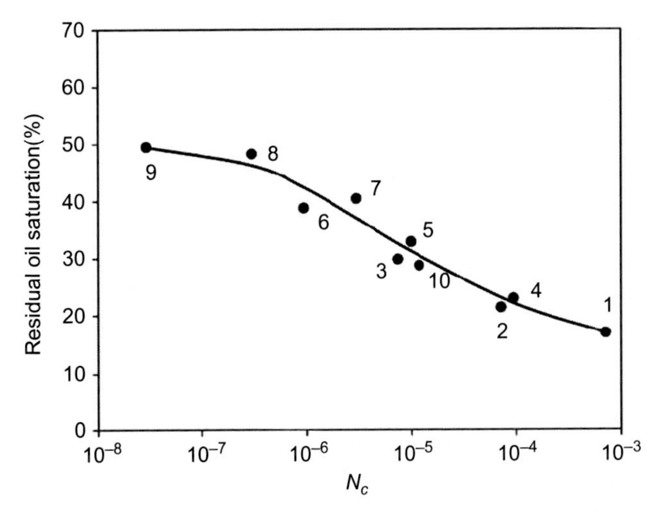

FIG. 7.1 Relationship curve between residual oil saturation and capillary number.

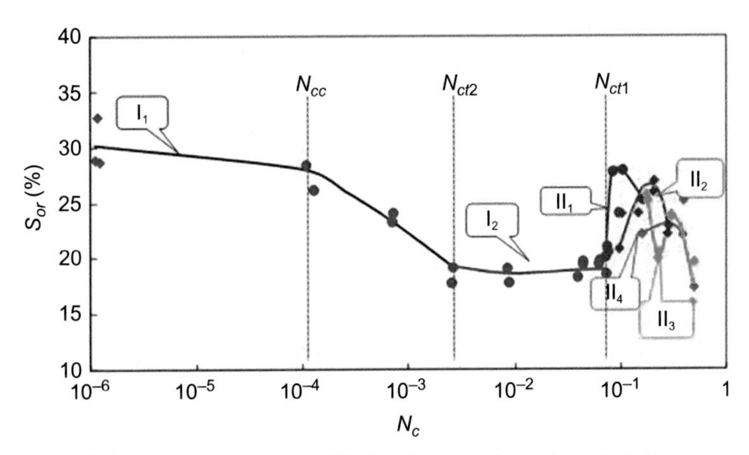

FIG. 7.2 Relationship experiment curve QL of capillary number and residual oil.

we can see that if QL is divided into two parts by limiting capillary number N_{ct2}, the left part looks like the classic capillary number experiment curve as shown in Fig. 7.1. The classic experimental curve of capillary number is similar in morphology, and the corresponding key capillary number value is similar as well. The right part of the figure shows there is a constant residual oil saturation S_{or}^{H} between the limit capillary number N_{ct2} and the limit capillary number N_{ct1}. When the limit capillary number is higher than the limit capillary number N_{ct1}, there are multiple capillary number curves, which change regularly.

7.1.2 The establishment of core microscopic oil/water distribution platform deepened the understanding of capillary number experiment curve

In Chapter 4 of this volume, the authors presented up-to-date research results and analysis of their experiment data. They also presented the core microscopic oil/water distribution platform according to thermodynamic law and the percolation mechanism. Fig. 7.3 shows the model sketch of experiment core microscopic spatial oil/water distribution characteristic model. In the descending order of aperture, the space is "pure oil" space (V_o), "pure water" space (V_w), "oil/water" coexistence space (V_{ow}, V_{wo}); the minimum may be the "pure oil" space or "pure water" space, which is determined by core wettability. "Pure oil" spatial displacement, the corresponding capillary number curve water drive section and water drive extension section enter the residual oil saturation sharply decreasing section with the capillary number N_{cc} as the turning point, At the limit of capillary number N_{ct2} turns into the "pure water" space V_w displacement, capillary number curve flat change, After the capillary number increased to greater than the limit capillary number N_{ct1}, it entered the "oil/water coexistence" space displacement, and the residual oil saturation was a composite function of capillary number, displacement velocity V, system viscosity μ_w interfacial tension σ_{ow} between displacement fluid and displaced fluid, corresponding to countless curves with regular changes in capillary number. The oil and water distribution model in the micro-space of experimental core is a technical platform for correct understanding of capillary number curve and the internal reasons for the complex shape of capillary number curve.

The experimental core microscopic spatial oil/water distribution model can be further extended to analyze and study displacement test results, which will deepen the understanding of displacement experiment study.

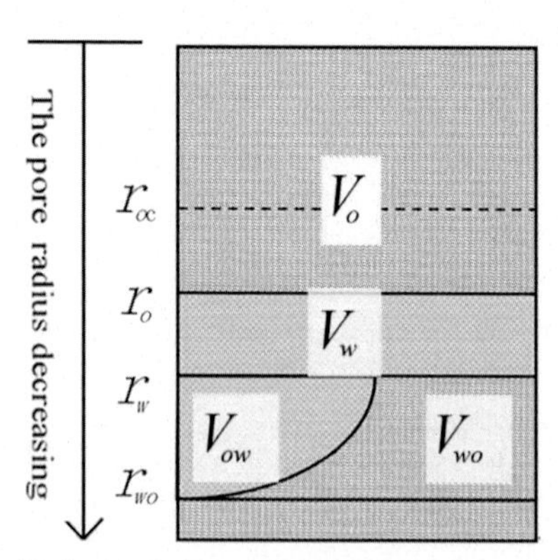

FIG. 7.3 The oil/water distribution in the core after the saturating process (oil-wet).

7.1.3 Phase permeability curve of compound displacement

As important as it is to understand that the experimental curve *QL* shows the relationship between capillary number and residual oil, it is also important to understand the change of relative permeability curve in the compound displacement process. Ref. 4 showed the mathematic description method and experiment determination key parameter method of the phase permeability curve in the compound displacement process.

It should be noted that the different residual oil saturation scope of phase permeability curve and the residual oil saturation of capillary number test curve *QL* correspond, and the following formula describes their corresponding relationship:

$$S_{or} = \begin{cases} S_{or}^{L} & N_c \le N_{cc} \\ S_{or}^{L} - \dfrac{N_c - N_{cc}}{N_{ct2} - N_{cc}}\left(S_{or}^{L} - S_{or}^{H}\right) & N_{cc} \le N_c \le N_{ct2} \\ S_{or}^{H} & N_{ct2} \le N_c \le N_{ct1} \\ S_{or}^{L}/(1 + T_o \times N_c) & N_c > N_{ct1} \end{cases} \quad (7.2)$$

By constructing the core micro oil/water distribution model, it becomes clearer that the capillary number in the drive oil process is different, that the oil/water flow pore space is different, corresponding to different relative permeability curves, and the description of the oil and water relative flow relationship is more accurate.

7.1.4 Research and application of software

A compound displacement software called Improved Mechanism of Combination Flooding Simulation (IMCFS) was developed based on relative permeability curve *QL* and corresponding capillary number curve *QL*. On the basis of a profound understanding of the capillary number theory, the software was used to study compound displacement via pilot tests and laboratory experiments.

The development of capillary number experimental curve *QL*, the core micro oil/water distribution model, and phase permeability curve of compound displacement are important advances in classic capillary number theory.

7.2 Research method of digitalized displacement experiments of compound displacement

Chapter 2 comprehensively summarized the research results of field tests in both China and the United States, and proposed a "digital oil displacement experiment."

7.2.1 Field experiments need more scientific and efficient research methods

Ref. 5 introduced the main results of the experiments conducted at Daqing Oilfield. At present, the "Type I" and "Type II" reservoirs in Daqing Oilfield are undergoing industrial application of strong-alkali ASP displacement. The industrial production has the following basic conditions: basic understanding of the theory of compound displacement technology, comprehensive understanding of industrial production oil reservoir conditions, and a mature approach for the optimization and design of compound displacement field production schemes

Field testing is an important and necessary research method for compound displacement technology. However, the cost of field testing is very high. The digital displacement oil experiment method is a necessary research method for understanding the field tests and improving their performance. Based on the summary of field tests, we can further optimize the displacement scheme and test it. Combining field test research and digital experimental research is a highly efficient method for studying the compound displacement technique.

Transitioning from "numerical simulation research" to "field test digital research" adds to the responsibility of numerical simulation workers and raises the demand for field test leaders. It is necessary to "understand the digital test" and take it as a "part of the field test."

7.2.2 The digital study of the compound displacement field test of Xingerxi block area in Daqing Oilfield

The compound displacement field experiment of Xingerxi block area in Daqing Oilfield began in 1996. The technical data of these experiments were obtained in Ref. 6. The sketch of the well site is detailed in Fig. 7.4. The IMCFS software was used to research the field experiments (see Chapter 2 for details of the research conditions).

The field test takes a five-spot pattern, four injection wells, nine production wells, including a central production wells, and a reservoir heterogeneous variation coefficient of 0.65. Ref. 7 proposed the simplified geological structure model describing different heterogeneous variation coefficients of oil reservoirs: reservoir areal homogenization, vertical heterogeneity three-formation structure, and different heterogeneous variation coefficient of oil reservoirs; the corresponding intervals have corresponding permeability. The different permutations of interval permeability determined the different sediment types of oil reservoirs. Fig. 7.5 shows the schematic diagram of the structure model. Studies confirmed that this model is suitable for the calculation and research of chemical oil displacement technology.

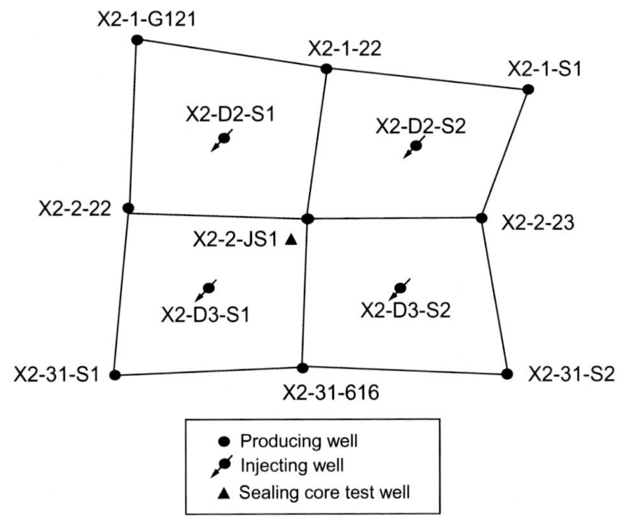

FIG. 7.4 Sketch of well location of ASP compound displacement test zone at Xingerxi block area.

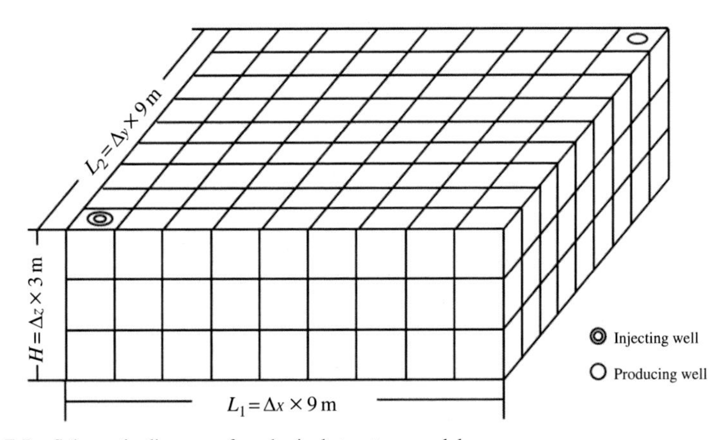

FIG. 7.5 Schematic diagram of geological structure model.

Because of the establishment of the core microscopic oil/water distribution model, the understanding of the geological model in the simulation study is deepened. The simplified geological structure model, which is based on the statistical method of multilevel permeability established by Dykstra and Parsons, is used in the fitting calculation of the field experiments. The model has the same heterogeneous variation coefficient as the heterogeneous reservoir. For the simplified geological structure mode, the three intervals are homogeneous cores with low, medium, and high permeability levels, which have no direct connection with the intervals of heterogeneous oil reservoir.

The oil displacement effect in this model is equivalent to that of heterogeneous reservoirs. Based on the microscopic oil/water distribution model, research can dig deeper into the microscopic space of the geological structural model, to study the oil/water flow changes in the microscopic space of each layer in the oil displacement process. This research method was adopted in Chapter 6 and is explained in detail here. The three-layer structure model used in the simulation calculation is compatible with the IMCFS software, which ensures the high accuracy of the fitting calculation results of the field test. According to the fitting calculation results, the micro model platform of oil layers is constructed. Table 7.1 shows that the parameters obtained from the platform not only contain data related to water displacement but also the limit capillary numbers N_{cc}, N_{ct2}, and N_{ct1}, the driving status conversion parameters T_1 and T_2, the initial oil content v_o in the pure oil pore space, and the oil content v_{ow} in the oil/water coexistence space, which is equivalent to the limit residual oil saturation S_{or}^H. Therefore the micro model provides precise parameters for researching compound displacement technology.

As can be seen from Ref. 7, fitting the water cut curve and oil production curve of the central well in Daqing Duanxi of block Beiyi polymer displacement test took much time and effort. The displacement experiment was carried out under the ultra-high water cut condition. Firstly, the authors fitted the displacement process and adjusted slightly the permeability of oil reservoir stratification and the related data of phase permeability curve layered residual oil saturation S_{or}^L of water displacement, and fitted the water displacement recovery target. The recovery percentage was 47.2% when the water cut of the produced fluid from oil well was 98%, which is consistent with the production data on the site. The basic physical parameters of the oil reservoir and the permeability of three intervals were corrected as 100×10^{-3}, 215×10^{-3}, and $525 \times 10^{-3}\,\mu m^2$. The authors continued to inject the water until the water cut of oil well reached 99.82% and the recovery reached 52.80%. Then, at the start of the compound displacement process, the authors converted to polymer preslug injection. The saturation parameter of residual oil S_{or}^H of the three intervals and the polymer solution viscosity-concentration curve shear rate both are amended to fit the curve between change of oil-well water cut and change of enhancement range of oil recovery. Fig. 7.6 displays the change curve of both water cut and recovery percentage of the oil wells between the on-site displacement experiment and fitting calculation, whose precision results were consistent. The fitting of displacement experiment achieved satisfactory results with high accuracy. In particular, the whole calculation process took only a few hours. The fitting calculation determined the oil reservoir geological data and corresponding information data of the compound displacement phase, established the digital geological model platform of the oil reservoir compound displacement data, which are needed to run the displacement scheme on the digital geological model platform. This is called the digital displacement experiment.

TABLE 7.1 Parameters of digitalized geological model for field experiments of Xingerxi in Daqing Oilfield.

	Permeability ($10^{-3}\mu m^2$)		Water and oil saturation (%)				Capillary number parameters				
Oil reservoir	*Before fitting*	*After fitting*	S_{wr}^L	S_{or}^L	v_o	v_{ow}	N_{cc}	N_{ct2}	N_{ct1}	T_1	T_2
Low permeability formation	100	100	24	36.5	60.5	15.5	0.0001	0.0025	0.0712	12	0
Medium permeability formation	250	215	22	34.5	63.5	14.5	0.0001	0.0025	0.0712	15	0
High permeability formation	725	525	21	32.5	65.5	13.5	0.0001	0.0025	0.0712	15	0

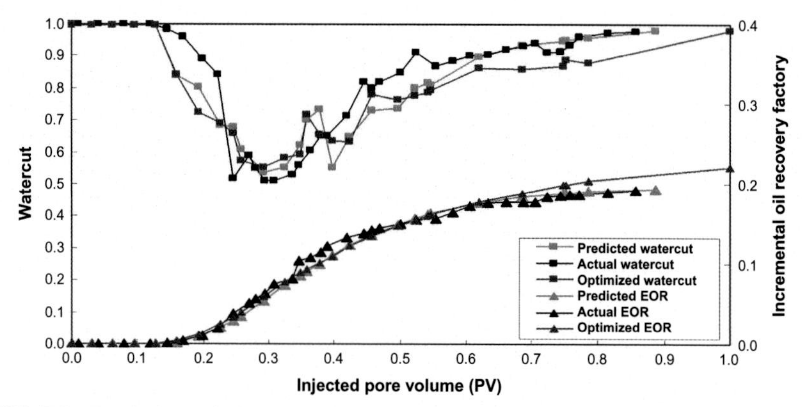

FIG. 7.6 Result curve of Xingerxi block area compound displacement experiment and plan.

7.2.3 Digital research of oil displacement experiments deepens the understanding of field tests

The digital displacement experiments provide rich ground and underground data. Based on data obtained from the digital displacement experiments, we have a clearer understanding of displacement and can obtain efficient and feasible displacement schemes so as to promote study and application of compound displacement technology.

The in-depth study was carried out via field experiments at Xingerxi block area on the digital geological model platform established by fitting the field experiment.

In the Xingerxi test reservoir, the interfacial tension of the system is at its lowest, 1.25×10^{-3} mn/m, and the viscosity of the system is at its highest, around 30 mPa s. Column 1 in Table 7.2 shows the areal grid capillary number of the high-permeability layer at Moment A when the compound displacement system is converted to 0.345 PV of oil reservoir in the experiment process. It can be learned that the capillary number in local range near the well grid is higher than the limit capillary number N_{ct1} of 0.0712, with the maximum value of 0.1687, placing it in a "Type II" displacement situation, corresponding with relatively high residual oil saturation. However, it is noted that the grid capillary number is in the range of 0.0025–0.0712 near the larger area, where it is in the best "Type I" displacement situation, corresponding with the limit value S_{or}^{H} of residual oil saturation for displacement. It is not difficult to understand that the "best" effect of a larger range is guaranteed exactly from the "overlimit" of the local range. Column 2 shows the high-permeability reservoir grid capillary number of 0.460 PV oil reservoir in-converting to compound displacement system injection at Moment B. It can be learned that there has been no grid whose number exceeds the limit capillary number N_{ct1} of 0.0712 in the whole reservoir, which indicates that the "Type II" displacement situation is the displacing

TABLE 7.2 Reservoir grid capillary number and residual oil saturation distribution table for different schemes of Xingerxi test at different time.

Item	Line/ Row	1	2	3	4	5	6	7	8	9
Capillary Number of high permeability layer at Moment A	1	0.1683	0.0883	0.0565	0.0403	0.0295	0.0195	0.0075	0	0
	2	0.0883	0.0686	0.0512	0.0386	0.0288	0.0192	0.0067	0	0
	3	0.0565	0.0512	0.0427	0.0343	0.0265	0.0176	0.0011	0	0
	4	0.0404	0.0386	0.0344	0.029	0.0223	0.0131	0	0	0
	5	0.0295	0.0289	0.0265	0.0223	0.0174	0.0005	0	0	0
	6	0.0196	0.0193	0.0177	0.0132	0.0005	0	0	0	0
	7	0.0076	0.007	0.0012	0	0	0	0	0	0
	8	0	0	0	0	0	0	0	0	0
	9	0	0	0	0	0	0	0	0	0
Capillary Number of high permeability layer at Moment B	1	0.0024	0.0279	0.039	0.0309	0.0253	0.0214	0.0158	0.0088	0.0011
	2	0.028	0.0425	0.0372	0.0305	0.0261	0.0224	0.0172	0.0111	0.0016
	3	0.0391	0.0373	0.0329	0.0291	0.0265	0.0233	0.0187	0.0136	0.0015
	4	0.031	0.0305	0.0291	0.0281	0.0265	0.0236	0.0199	0.0155	0.0009
	5	0.0254	0.0261	0.0266	0.0265	0.0254	0.0233	0.0208	0.0169	0.0003
	6	0.0214	0.0225	0.0234	0.0237	0.0233	0.0226	0.0208	0.0195	0.0001
	7	0.0159	0.0172	0.0187	0.0199	0.0207	0.0208	0.0213	0.0097	0
	8	0.0088	0.0111	0.0136	0.0154	0.0165	0.0203	0.0119	0.0025	0
	9	0.0011	0.0016	0.0016	0.0009	0.0003	0.0001	0	0.0001	0.0001
Remaining Oil of High-permeability Reservoir at Termination Moment (%)	1	4.7	0.4	0.6	0.7	0.6	0.5	0.5	0.8	3.3
	2	0.4	0.5	0.7	0.7	0.5	0.4	0.5	0.7	2.4
	3	0.6	0.7	0.8	0.6	0.4	0.4	0.5	0.7	2.7
	4	0.7	0.7	0.6	0.4	0.4	0.4	0.5	0.7	2.7
	5	0.6	0.5	0.4	0.4	0.4	0.4	0.5	0.7	2.6
	6	0.5	0.4	0.4	0.4	0.4	0.4	0.5	0.7	2.4
	7	0.5	0.5	0.5	0.5	0.5	0.5	0.5	0.6	2.1
	8	0.7	0.7	0.7	0.7	0.7	0.6	0.6	0.6	1.7
	9	3.2	2.4	2.6	2.6	2.5	2.2	2.9	1.5	1.9
Remaining Oil of Low-permeability Reservoir at Termination Moment (%)	1	0.2	0.3	0.5	0.7	1.3	3.5	**18.7**	**24.1**	**35.1**
	2	0.3	0.4	0.5	0.8	1.4	4.2	**17.3**	**24.4**	**34.8**
	3	0.5	0.5	0.7	1.0	1.6	5.1	**17.0**	**25.0**	**34.1**
	4	0.7	0.8	1.0	1.3	2.0	7.8	**18.1**	**25.9**	**34.0**
	5	1.3	1.4	1.6	2.0	4.3	**10.9**	**20.4**	**26.9**	**33.7**
	6	3.7	4.0	4.9	7.7	**10.7**	**16.5**	**24.8**	**28.2**	**33.4**
	7	**18.6**	**17.1**	**16.9**	**18.0**	**20.2**	**24.6**	**28.2**	**28.1**	**33.1**
	8	**24.1**	**24.3**	**24.9**	**25.8**	**26.9**	**28.1**	**28.0**	**28.5**	**31.8**
	9	**35.0**	**34.8**	**34.2**	**34.0**	**33.7**	**33.3**	**33.1**	**31.8**	**32.3**
Surfactant Concentration of Low-permeability Reservoir at Termination Moment (%)	1	0.001	0.002	0.01	0.03	0.063	0.115	0.131	0.113	0.037
	2	0.002	0.005	0.017	0.032	0.073	0.116	0.132	0.117	0.053
	3	0.01	0.018	0.031	0.043	0.089	0.122	0.133	0.121	0.061
	4	0.031	0.032	0.045	0.079	0.099	0.116	0.125	0.123	0.066
	5	0.06	0.073	0.087	0.1	0.098	0.09	0.113	0.125	0.071
	6	0.11	0.117	0.119	0.115	0.088	0.085	0.103	0.124	0.079
	7	0.131	0.132	0.128	0.12	0.107	0.098	0.112	0.122	0.08
	8	0.114	0.118	0.12	0.123	0.124	0.12	0.119	0.116	0.082
	9	0.038	0.056	0.064	0.071	0.074	0.079	0.082	0.081	0.065

process of a relatively short time. The vast majority of the grid capillary numbers in the whole reservoir are 0.0025–0.0712, which is in the best "Type I" displacement situation. With the concept of oil and water distribution in microscopic space, the change of residual oil content can be studied by analyzing the flowing change of oil content v_o in microscopic pure oil space V_o or oil content v_{ow} in oil and water coexistence space V_{ow}.

To facilitate the following comparative analysis, first we analyze the low-permeability layer. In the process of oil displacement, because the grid capillary number N_c of the low-permeability layer has never exceeded the limit capillary number N_{ct1}, the oil in V_{ow} has not been driven away, and oil content

v_{ow} remains unchanged at the initial value of 15.5%, which corresponds to the residual oil saturation limit S_{or}^{H}. When the displacement process is terminated, the produced oil comes from the crude oil stored in space V_o and the grid residual oil saturation S_{or} contains the oil content v_{ow} that is not driven out in the space V_{ow}, and the retention oil in the space V_o and V_w, which is still marked as v_o, and $v_o = S_{or} - v_{ow}$. Column 4 in Table 7.2 shows the residual oil content v_{ow} distribution data of the low-permeability reservoir grid after the end of the displacement experiment. The residual oil content v_{ow} is not listed because it remains constant at 15.5% in the space V_{ow}, but the oil content distributed in others spaces is given in red numbers on the grid. It can be seen that there is relatively low oil saturation v_o only on nearly half a level in the side of the water well in the upper left corner. There is high oil saturation on the grid in a large area in the front, where the "oil wall" is gathered here by the crude oil flooded from the rear area. The research shows the targets for people to further tap—residual oil v_o of low-permeability reservoirs. Column 5 in Table 7.2 shows the surfactant content distribution data of the low-permeability reservoir grid after the end of the displacement experiment. According to the corresponding surfactant concentration distribution, it can be seen clearly that the position of the "oil wall" corresponded to high surfactant concentration, where the main body of the composite system slug is finally resided. These results also tell us that what should be done to enhance oil recovery is to further advance the compound system solution that entered the low-permeability reservoir. To achieve the targets, the viscosity of the polymer slug behind the main body slug should not be less than the viscosity of the main body slug, and the slug must have enough length. Only through this way can the water phase intrusion in the high-permeability reservoir be inhibited to maintain the good driving momentum and ensure more oil to be produced from the low-permeability reservoirs so as to obtain a better displacement effect.

Chapter 6 carefully analyzed the residual oil distribution in the high-permeability grid; corresponding data is presented in Column 3 in Table 7.2. The displacement situation on the adjacent side of the water well is under the condition of a greater than the limit capillary number N_{ct1} for a short period in the displacement process. The residual oil distribution data clearly shows the influence effect of the displacement situation. In the well grid, the residual oil saturation S_{or} is only 4.7%, far less than the limit residual oil saturation S_{or}^{H} of 13.5%. In this case, the value of S_{or} contains only the grid residual oil content v_{ow}. The oil in the space V_o and V_w has been entirely driven out, so the grid residual oil content v_o is zero. In the areas outside of the water well grid marked in light green, all the oil in the space V_o and V_w has been driven out, and the grid residual oil content is also zero. The value of S_{or}^{H} originally stored in the space V_{ow} does not decrease, but increases the amount of oil captured; the number in the grid is exactly the corresponding increment, which is around 0.7%. The area marked in orange is the "fuzzy" area of residual oil content in the microscopic space; the numbers of oil content in the grid contain not only the remaining oil trapped in space V_o and V_w space, but also the flowing oil captured in the grid space V_{ow}. Close to the sides of the oil well, the number marked in red is the grid residual content v_o, which is usually less, and the maximum residual content of the corner grid is 3.3%, but 1.9% of the oil well grid. In this case, the driven situation is under good condition in grid space V_o, but the oil in the grid space V_{ow} is not driven and the oil content v_{ow} remains unchanged at the initial value of 13.5%.

Now, let's comprehensively analyze the technological and economical effects of the experiments. For the low- and middle-permeability oil layers, the displacement process is not under the "Type II" displacement situation. The oil content in the space V_{ow} of two layers keeps the initial values of 15.5% and 14.5% respectively, but the oil stored originally in the pure oil space V_o is partly produced, and the remaining oil is left in space V_o and V_w. For the low-permeability layer, the original oil content v_o is 60.5% and the produced oil is 44.08%, which is 72.86% of the total. The remaining oil is 16.42%, which is 27.14% of the total. For the middle-permeability layer, 63.5%, 60.04%, 94.55%, 3.46%, 5.45% respectively. For the high-permeability oil layers, the displacement process is only under the "Type II" displacement situation within a local scope in a relatively short time, resulting in the remaining oil content in the oil/water spaces varying to a certain extent, therefore it cannot be accurately measured. In this case, it is still considered that the residual oil content in the oil/water space V_{ow} retains the initial value of 13.5%, and the produced oil all comes from the pure oil space V_o. Further, we can obtain that the original oil content v_o is 65.5% and the produced oil is 64.68%, which is 98.75% of the total. The remaining oil is 0.82%, which is 1.25% of the total. In the three layers, especially the middle- and high-permeability layers, a large amount of oil is produced from the pure oil space, ensuring the test has a relatively high recovery ratio and high enhanced oil recovery range. By calculation, we have obtained the incremental oil for per ton equivalent polymer of 60.13 tons as the surfactant price is 1.5 times the polymer price. From an economical point of view, this is a fairly ideal result.

7.3 Optimized research of the industrial application scheme of compound displacement in Daqing Oilfield

In order to satisfy the industrial application of complication displacement at Daqing Oilfield, several important issues are subject to in-depth studies.

7.3.1 Further optimization of the field experiment oil displacement scheme

The industrial experiment of Duandong in Beiyi block of No. 1 Oil Production Plant of Daqing Oilfield was a huge success, with an overall enhanced recovery of more than 25% and an enhanced recovery of 30% for high-production well groups. Further optimization and improvement is carried out to the displacement schemes based on the Duandong experiment.

A five-spot flood well system was used for the Duandong industrial experiment, with injector-producer spacing of 125 m. The experiment results verified that with this kind of well pattern, well spacing is worthy of promotion and application. The field schemes were gradually improved to calculate different comparative schemes. Table 7.3 shows the field experiment scheme and the gradually improved digital displacement scheme. Table 7.4 shows the calculation results of corresponding displacement schemes.

Experiment Scheme: 0.054 PV polymer advanced profiling slug, 0.351 PV ASP system master slug, NaOH concentration 1.2%; afterwards, the alkali concentration for slave slug of two-stage compound system decreased to 1.0%, the surfactant concentration of the three-stage slug was 0.2%, and polymer protective slug volume was 0.2 PV with a concentration of 1000 mg/L. Table 7.4 shows that the final recovery percentage was 69.91%, the water-flood enhancement of recovery was 26.24%, and the oil increment for per ton equivalent polymer was 55.63 tons.

Adjustment Scheme 1: It is generally believed that the compound displacement scheme should have a large compound system slug, just as in this experiment scheme; the compound system slug volume is 0.3 PV or so in the optimization scheme of digital displacement experiment studies. Initial experiments: Secondary Slug 2 of compound system is changed into a polymer slug. The calculation results shows that the recovery percentage of the scheme is 69.34% and the enhancement oil recovery is 25.67%, which is slightly lower than the experiment scheme. Compared to the experiment scheme, the surfactant dosage decrease by 58.9 tons, and the oil increment for per ton equivalent polymer increased to 67.07 tons, which is 11.44 tons more than the experiment scheme.

Adjustment Scheme 2: The subsequent polymer slug of the compound system slug is usually called the protective slug, which is separated from the subsequent freshwater slug to prevent water onrush and protect the compound system slug to move forward. This is a kind of passive protection mode. The positive mode refers to the fact that the viscosity of the subsequent polymer slug is not less than that of the compound system slug and that there is a sufficiently big volume in order to continue to expand the swept effects and displace more compound system solution to the low-permeability horizons and to two wings of the mainstream line. This is done to achieve better displacement effects. On the basis of Adjustment Scheme 1, the concentration of the subsequent polymer slug of the compound system slug is increased to 1900 mg/L. The calculation results shows that the residual oil value of the three intervals is all less than the corresponding value of the experiment scheme at the termination of the displacement process. The recovery percentage of the scheme is 70.61%, which is greater than the experiment scheme. Compared to the experiment scheme, the surfactant consumption decreases and the polymer consumption increases.

TABLE 7.3 Further optimization of industrial experiment scheme of combination flooding in Duandong (east of fault) of Beiyi block.

	Preslug		Master slug of combinational system			Slave Slug 1 of combinational system			Slave Slug 2 of combinational system			Polymer slug	
Scheme	*P concentration (mg/L)*	*Volume (PV)*	*S degree (%)*	*P concentration (mg/L)*	*Volume (PV)*	*S concentration (%)*	*P concentration (mg/L)*	*Volume (PV)*	*S concentration (%)*	*P concentration (mg/L)*	*Volume (PV)*	*P concentration (mg/L)*	*Volume (PV)*
Experiment	1300	0.054	0.2	1900	0.351	0.2	1900	0.078	0.2	1650	0.15	1000	0.20
Adjustment 1	1300	0.054	0.2	1900	0.351	0.2	1900	0.078		1000	0.15	1000	0.20
Adjustment 2	1300	0.054	0.2	1900	0.351	0.2	1900	0.078		1900	0.15	1900	0.20
Adjustment 3	1300	0.054	0.2	1900	0.3		1900	0.129		1900	0.15	1900	0.20
Adjustment 4	1300	0.054	0.3	1900	0.3		1900	0.129		1900	0.15	1900	0.20
Adjustment 5	1300	0.054	0.3	1900	0.3		1900	0.129		1900	0.15	1900	0.20
Adjustment 6	1300	0.054	0.3	2400	0.3		2400	0.129		2400	0.15	2400	0.20
Adjustment 7			0.3	2400	0.3		2400	0.172		2400	0.15	2400	0.20
Adjustment 8			0.3	2325	0.30		2325	0.250		2325	0.15	2325	0.20
Adjustment 9	1300	0.054	0.3	2325	0.30		2325	0.196		2325	0.15	2325	0.20

The combination flooding slug in the table refers to the strong-base ASP system, and NaOH concentration is 1.2%–1.0%.

TABLE 7.4 Flooding effect of further optimization of industrial experiment scheme of combination flooding in east of fault of North-1 Block of No. 1 Oil Production Company of Daqing Oilfield.

Scheme	Implementation time (days)	Reservoir maximum average pressure ratio	Slug polymer concentration (mg/L)	Maximum viscosity of displacing phase (mPa s)	Recovery percent of reserves (%)	Enhancement of recovery (%)	Stratified residual oil, S_{or} (%)			Surfactant consumption (t)	Total chemical dosage equivalent polymer (t)	Oil increase for T equivalent polymer (t/t)
							Upper layer	*Middle layer*	*Lower layer*			
Experiment	619	0.9617	1900	23.26	69.91	26.24	36.49	19.71	14.06	224.6	602.5	55.63
Adjustment 1	647	0.9617	1900	23.18	69.34	25.67	36.76	20.55	14.28	166.3	488.9	67.07
Adjustment 2	675	0.9617	1900	24.18	70.61	26.94	36.19	18.57	13.87	166.3	549.9	62.58
Adjustment 3	630	0.9650	1900	23.95	69.82	26.15	36.98	19.53	13.95	116.5	475.2	70.30
Adjustment 4	653	0.9315	1900	23.98	71.07	27.40	35.68	17.99	13.91	174.7	562.5	62.22
Adjustment 5	970	0.7332	1900	24.63	71.34	27.67	34.54	18.38	13.98	174.8	562.7	62.81
Adjustment 6	980	0.9551	2400	34.57	74.12	30.45	30.72	16.20	13.50	174.8	638.1	60.96
Adjustment 7	1010	0.9883	2400	34.50	74.32	30.65	30.21	16.12	13.54	174.8	637.8	61.39
Adjustment 8	1042	0.9936	2325	33.07	74.86	31.19	29.08	16.0	13.55	174.8	668.1	59.64
Adjustment 9	1043	1.0052	2325	33.11	74.67	31.00	29.58	16.05	13.51	174.8	667.1	59.36

However, the total expense of chemicals reduced and the per ton equivalent polymer increases oil by 6.95 tons compared with the experiment scheme.

Adjustment Scheme 3: The master slug of the compound system is changed to 0.3 PV. The calculation results shows that compared to the experiment scheme, the surfactant consumption in the scheme relatively reduces by nearly half, the polymer consumption increases by 1/5 or so, the total expense of chemicals decreases by 1/4, and the final recovery percentage decreases by 0.09% compared to the experiment scheme. The per ton equivalent polymer increases oil to 70.30 tons. It is noted that the residual oil content of the high- and middle-permeability layers is less than the experiment scheme, but it is slightly higher than the experiment scheme for the low- permeability layer when the displacement process is terminated.

Adjustment Scheme 4: The surfactant concentration of the compound system slug in Adjustment Scheme 3 is raised to 0.3%. The recovery percentage in the scheme increases to 71.07%, which is an increase of 1.16% compared to the experiment scheme. The residual oil of the three intervals reduces to 13.91%, 17.99%, and 35.68%, respectively, which is a decrease of 0.15%, 1.72%, and 0.81%, respectively. The surfactant consumption relatively decreases by 30%, and the oil increase for ton equivalent polymer is 62.22 tons, which is an increase of 6.59 tons compared to the experiment scheme. It can be learned by researching the change of surfactant concentration field that the high concentration system surfactant has a relatively good diffusion effect in the oil reservoir, which is conducive to expanding the spread range and enhancing the displacement effect.

Adjustment Scheme 5: According to Chapter 2, the seepage velocity and system viscosity have different contributions in the displacement process. The contribution of seepage velocity only contained the displacing action, however, the enhancement of system viscosity can not only inhibit the displacing fluid onrush to expand the spread effect and improve the displacement effect of two wings of reservoir plane mainstream line and low-permeability reservoir, but also improve oil/water mobility ratio, which is conducive to the mobility and production of the oil. It had outstanding contributions to the enhancement of the displacement effect. Therefore, the measure of "Deceleration and Viscosity Stabilization" is proposed. It is recommended to use injection velocity of 0.40 PV/y while the injector-producer spacing is 125 m. The decelerated injection is applied to all subsequent slugs of the compound system slug in Adjustment Scheme 4 and the system viscosity is not changed. Compared to the previous scheme, the consumption of chemicals is basically the same, the maximum viscosity of the underground system increases slightly, which is related to the weakening of shearing and degradation action, the stratified residual oil and upper low-permeability reservoir decreases, which is related to the relative increase of system viscosity, and the high-permeability reservoir in the middle and lower oil reservoirs relatively increased, which is related to the decrease of injection velocity. Compared to Adjustment Scheme 4, the recovery percentage of this scheme only decreased by 0.27%, and the oil increase for ton equivalent polymer reduced 0.59 tons. It is noted in particular that the maximum average pressure of the oil reservoir decreases significantly, which creates the conditions for further adjusting and enhancing the system viscosity.

Adjustment Scheme 6: Compared to Adjustment Scheme 5, the polymer concentration in the compound system slug and subsequent polymer slug is increased to 232 4 mg/L. Polymer consumption increases by about 25%, the final recovery percentage increases by 3.05%, and the oil increase for ton equivalent polymer is 60.96 tons, compared with Adjustment Scheme 4. We can clearly see the significant effect of "Deceleration and Viscosity Stabilization."

Adjustment Scheme 7: The polymer preslugs are added in front of the compound system slug for the purpose of "increasing the entry of compound system into the low permeability reservoir to improve the displacement effect because the polymer pre-slug has the 'profile control' effect." However, there is no such slug structure design in the United States. Remove the "preslug" in Adjustment Scheme 5 and apply the saved polymer to the subsequent "protective" slug and then calculate the adjustment scheme. The table shows that the consumption of chemicals in the two comparative schemes is the same. The recovery percentage of the latter scheme is slightly better than that of the previous scheme; low/middle-permeability reservoir of stratified residual oil slightly decreases and the high-permeability reservoir slightly increases. The results show that the effect of "profile control" of the polymer preslug is not obvious.

Adjustment Scheme 8: The subsequent slug of large volume and high concentration polymer is beneficial to the longitudinal improvement of low-permeability model displacement and the expansion of the spread of the two wings of the main line on the plane, which has an obvious "profile control" effect and improves oil displacement. Based on the studies in Chapter 2, it is recommended to use subsequent polymer slug volume of 0.6 PV to calculate the scheme. The calculation results shows that the maximum average pressure of the oil reservoir is the same as the experiment of Xingerxi block area, meeting the requirements of injection condition. The implementation time of the scheme is 1042 days, meeting the stability requirements. The oil increase for ton equivalent polymer is 59.64 tons, meeting the requirements of economic indicators. The final recovery percentage of the scheme is 74.86%, an increase of 31.19% compared to water displacement, and 4.95% compared to the experiment scheme.

Adjustment Scheme 9: In order to further assess the effect of the preslug, we reduced the volume of the subsequent polymer slug and added the preslug to calculate the scheme on the basis of Adjustment Scheme 8. The consumption of chemicals was the same when comparing the two schemes. Compared to the former scheme, the recovery percentage of the latter scheme decreased by 0.19%, the residual oil of the low- and middle-permeability layers increases to 0.5% and 0.05% respectively, and residual oil of the high-permeability layer decreased to 0.04%, which verified again that it is unnecessary to set up the preslug.

The gradual adjustments of the schemes determined the optimization of the displacement scheme and established the gradual optimization adjustment method for field schemes.

7.3.2 Impact of interfacial tension changes on displacement effect

The compound displacement technology adopted the weak-base system in the field experiments in the United States. In Ref. 5, Daqing Oilfield attached great importance to the selection of alkali in the system. The weak-base Na_2CO_3 system was selected in the experiment of PO Block of No. 1 Oil Production Plant in 1993, and the oil recovery increased by 21.4% compared to water displacement. The strong-base NaOH system was used in the experiment of Xingwu block of No. 4 Oil Production Plant in 1994, and the oil recovery relatively increased by 25%. Afterwards, most of the experiments used the strong-base NaOH system although the weak-base Na_2CO_3 system was not abandoned. In the industrial application experiments, the strong-base system was used in experiments of Xinger block of No. 4 Oil Production Plant, Nanwu block of No. 2 Oil Production Plant, and Duandong in Beiyi block of No. 1 Oil Production Plant. The weak-base system was used in the experiments of West Nansan block of No. 3 Oil Production Plant. The results of field experiments in No. 3 Oil Production Plant show that the weak-base system does relatively small "harm" to the reservoir and has a good oil displacement effect. However, under the same experiment conditions in the oil/water circumstance in Daqing Oilfield, the strong-base system can obtain lower interfacial tension to achieve a better displacement effect. The good technical and economic results can be used to make the investment on reducing the "harm" of the base/alkali.

The studies of capillary number curve show that seepage velocity, system interfacial tension, and system viscosity are three factors that decided the size of capillary number. Under the certain condition of seepage velocity and relatively high system viscosity, the lower the system oil/water interfacial tension is, the bigger the capillary number will be in the displacement process and the better the displacement effect will be. Based on the digital geology model of Duandong of Beiyi block of No. 1 Oil Production Plant and the optimization scheme in Table 7.4, and the maximum average pressure of oil reservoir in injection process of Scheme 6 is treated as the control conditions to calculate the different displacement experiment schemes with different interfacial tension. The calculation results are listed in Table 7.5.

Table 7.5 shows that the implementation time of all displacement schemes is within 3 years, meeting the stability requirements. With the decrease of system interfacial tension, the maximum capillary number of the grid in the displacement process increases as does the recovery percentage of the scheme. The schemes can be organized into four groups.

Scheme 1 is the high interfacial tension scheme with interfacial tension of 0.025 mN/m. The maximum capillary number of the grid in the displacement

TABLE 7.5 Flooding effect of different system interfacial tensions for industrial experiment schemes of combination flooding in Duandong (east of fault) of Beiyi block.

Scheme	Minimum interfacial tension (10^{-3} mN/m)	Implementation Time (d)	Reservoir maximum average pressure ratio	Maximum viscosity of displacing phase (mPa·s)	Maximum capillary number of the grid in flooding Process	Recovery percent of reserves (%)	Enhancement of recovery percent (%)	Stratified Residual Oil S_{or} (%)			Total chemical dosage equivalent polymer (t)	Oil increase for T equivalent polymer (t/t)
								Low permeability layer	Medium permeability layer	High permeability layer		
1	25.0	911	1.00	34.44	0.015	62.96	19.26	37.4	29.8	19.1	678.2	36.28
2	10.0	977	0.99	32.67	0.036	69.32	25.62	34.6	22.4	14.5	661.6	49.47
3	7.50	999	1.00	32.36	0.048	70.94	27.24	33.8	19.8	14.1	658.1	52.87
4	5.00	1025	0.99	32.43	0.072	72.53	28.83	32.4	17.7	13.9	658.1	55.96
5	2.50	1038	0.98	32.60	0.146	74.13	30.43	30.2	16.4	13.7	661.6	58.75
6	1.25	1045	1.00	33.03	0.297	74.88	31.18	29.0	16.0	13.5	665.8	59.82
7	0.75	1043	1.00	32.74	0.499	75.06	31.36	28.7	15.9	13.6	667.7	59.99
8	0.50	1042	1.01	33.27	0.75	75.29	31.59	28.5	15.9	13.2	667.7	60.42
9	0.25	1044	1.00	33.13	1.49	75.99	32.29	28.4	16.2	11.8	667.7	61.77
10	0.02	1117	1.01	36.21	18.94	78.00	34.30	28.5	13.3	8.47	**852**	**51.42**

process is 0.015, the recovery percentage is 62.96%, the oil recovery range is 19.28%, and the oil increase for ton equivalent polymer is 36.28 tons. The technical and economic results are very poor.

In Schemes 2–4 the interfacial tension ranges from 0.005 to 0.01 mN/m, which can be considered a weak-base system. The calculation results show that the maximum capillary number range of the grid in the displacement process is less than 0.072, all three schemes are in the condition of "Type I" displacement, and the oil in the oil/water coexistence space has not been activated. The residual oil content S_{or} of every interval of the three schemes is listed in Table 7.5. The residual oil content of Scheme 2 in the high-, middle-, and low-permeability layers is 14.5%, 22.4%, and 34.6%, respectively. As the interfacial tension decreases, the residual oil value in each layer decreases. It is clear that Scheme 4 with the lowest interfacial tension has the best oil displacement effect with a recovery percentage of 72.53%, an increased oil recovery of 28.83%, and an increase of oil for ton equivalent polymer of 55.96 tons. Table 7.6 shows the residual oil content v_o distribution data of the 1/4 well group grids of the high-permeability layer at the termination moment in Scheme 4; the value is relatively small whose maximum value is 6.39% in the corners of two wings of mainstream line, and 1.30% of oil well grids. In this case, it shows that the oil layers have not experienced "Type II" driving situation, and the residual oil content v_{ow} in the microscopic oil/water co-existence space remains at 13.2%.

In Schemes 5–8, the system interfacial tension ranges from 0.0005 to 0.0025 mN/m, which can be considered a strong-base system. The interfacial tension of Scheme 6 is 1.25×10^{-3} mN/m, 1/4 of Scheme 4, and the maximum grid capillary number of water well is 0.297, which is four times that of Scheme 4.

TABLE 7.6 Residual oil distribution of high-permeability layer grid at different termination moments (%).

scheme	Line/ row	1	2	3	4	5	6	7	8	9
	1	0.04	0.07	0.13	0.19	0.25	0.32	0.4	0.55	6.39
	2	0.07	0.1	0.15	0.2	0.25	0.32	0.39	0.51	1.49
	3	0.13	0.15	0.18	0.23	0.26	0.32	0.39	0.47	1.47
scheme	4	0.19	0.2	0.23	0.25	0.28	0.35	0.37	0.46	1.62
	5	0.25	0.25	0.26	0.28	0.33	0.35	0.37	0.41	1.62
4	6	0.32	0.32	0.32	0.34	0.35	0.37	0.37	0.41	1.56
	7	0.39	0.39	0.38	0.37	0.37	0.36	0.32	0.37	1.46
	8	0.53	0.5	0.46	0.46	0.42	0.41	0.35	0.29	1.09
	9	6.01	1.41	1.45	1.60	1.58	1.48	1.33	0.88	1.30
	1	7.465	12.04	0.18	0.22	0.27	0.33	0.39	0.48	1.15
	2	12.04	0.55	0.32	0.27	0.29	0.33	0.38	0.46	1.11
	3	0.18	0.33	0.29	0.29	0.30	0.33	0.38	0.44	1.16
scheme	4	0.22	0.27	0.29	0.30	0.31	0.34	0.38	0.43	1.18
	5	0.27	0.29	0.30	0.31	0.33	0.34	0.36	0.40	1.15
6	6	0.33	0.33	0.33	0.34	0.34	0.34	0.33	0.36	1.11
	7	0.39	0.38	0.38	0.37	0.36	0.32	0.29	0.31	1.05
	8	0.48	0.46	0.44	0.43	0.41	0.36	0.29	0.26	0.83
	9	1.12	1.08	1.12	1.15	1.13	1.05	0.94	0.66	1.00
	1	9.47	11.30	12.98	0.65	0.57	0.52	0.49	0.53	1.13
	2	11.30	12.29	0.06	0.61	0.55	0.51	0.49	0.51	1.13
	3	12.97	0.06	0.64	0.59	0.53	0.49	0.48	0.49	1.16
scheme	4	0.65	0.61	0.59	0.54	0.50	0.48	0.46	0.47	1.19
	5	0.57	0.55	0.52	0.51	0.50	0.46	0.44	0.45	1.16
7	6	0.52	0.51	0.49	0.49	0.46	0.45	0.39	0.38	1.12
	7	0.49	0.48	0.48	0.46	0.44	0.39	0.35	0.34	1.06
	8	0.53	0.51	0.49	0.47	0.46	0.38	0.32	0.27	0.76
	9	1.11	1.09	1.13	1.16	1.14	1.05	0.89	0.39	12.54
	1	8.72	10.24	11.63	12.90	0.82	0.61	0.55	0.57	1.13
	2	10.24	11.09	11.87	12.98	0.78	0.59	0.54	0.55	1.14
	3	11.63	11.87	12.58	0.20	0.76	0.67	0.54	0.52	1.17
scheme	4	12.88	12.98	0.19	0.64	0.76	0.78	0.55	0.51	1.30
	5	0.81	0.77	0.77	0.77	0.77	0.71	0.57	0.51	1.37
8	6	0.60	0.59	0.76	0.79	0.70	0.60	0.52	0.46	0.85
	7	0.55	0.55	0.55	0.56	0.58	0.51	0.43	13.11	12.77
	8	0.57	0.54	0.51	0.51	0.51	0.42	13.01	11.49	12.28
	9	1.11	1.11	1.34	1.27	1.25	0.42	12.29	11.79	11.81
	1	8.00	9.44	10.23	10.96	11.90	12.92	0.96	0.76	1.23
	2	9.44	10.15	10.08	10.82	11.74	12.59	0.47	0.94	1.38
	3	10.22	10.08	10.51	11.09	11.68	12.29	13.10	0.56	0.99
scheme	4	10.92	10.80	11.08	11.40	11.68	12.05	12.44	12.58	13.16
	5	11.90	11.74	11.65	11.66	11.74	11.78	11.83	11.81	12.59
9	6	12.88	12.57	12.26	12.01	11.76	11.54	11.25	11.01	11.90
	7	0.94	0.43	13.05	12.38	11.77	11.23	10.73	10.30	11.33
	8	0.75	0.93	0.50	12.52	11.74	10.92	10.25	9.55	11.00
	9	1.21	1.35	0.83	13.02	12.42	11.64	11.18	10.86	10.77
	1	7.08	8.23	8.09	7.81	7.94	8.08	8.32	9.03	10.96
	2	8.23	8.63	7.96	7.68	7.83	7.96	8.17	8.55	9.88
	3	8.10	7.98	7.59	7.54	7.68	7.87	7.95	8.22	9.64
scheme	4	7.82	7.67	7.54	7.58	7.57	7.67	7.77	8.00	9.63
	5	7.94	7.83	7.68	7.56	7.63	7.61	7.65	7.83	9.59
10	6	8.06	7.96	7.86	7.71	7.61	7.56	7.55	7.64	9.30
	7	8.33	8.17	7.94	7.76	7.64	7.54	7.49	7.48	9.05
	8	9.02	8.53	8.21	7.99	7.81	7.63	7.50	7.43	8.84
	9	10.89	9.82	9.57	9.54	9.52	9.30	9.05	9.30	9.23

It has appeared a "Type II" driving situation in displacement process. From the residual oil distribution data of the high-permeability layer listed in Table 7.6, we conclude that the displacement process was the same as that of the experiment at Xingerxi block area with a limited range of the "Type II" driving situation and in a relatively short time. The "best" flooding effect of a larger range was guaranteed exactly from the "overlimit" of local range, and the oil recovery is improved compared to Scheme 4. As seen in Table 7.5, the residual oil in the three layers (low, medium, and high permeability) decreased by 3.4%, 1.7%, and 0.4% respectively. From Table 7.6, we can see that the residual oil saturation significantly reduced in the locations of the edges and corners of the two wings of the mainstream line, and the effect of oil recovery increased. The enhancement oil recovery increased by 31.18%, with is 2.35% higher than Scheme 4. In Scheme 5, the interfacial tension is 2.5×10^{-3} mN/m, which is twice that of Scheme 4. The maximum capillary number of well grid in the oil displacement process is 0.146, which doubles that of Scheme 4, and is half of Scheme 6. In this case, the influence of "Type II" driving situation is weaker than in Scheme 6, the residual oil S_{or} in the three layers compared to Scheme 4, and increases by 1.2%, 0.4%, and 0.2% respectively compared to Scheme 6. The enhancement oil recovery is 30.43%, 1.6% greater than in Scheme 4 and 0.75% less than in Scheme 6. The oil increase for ton equivalent polymer is 58.7 tons, an increase of 2.79 tons compared to Scheme 4 and a decrease of 1.07 tons compared to Scheme 6. In Scheme 7, the interfacial tension is 7.5×10^{-4} mN/m and the maximum capillary number of well grid in the oil displacement process is 0.499, which is a great improvement compared with Scheme 6. Table 7.6 shows that the scope and extent influenced by the "Type II" driving situation relatively increased in the displacement process. From Table 7.5, we can see that the residual oil content in the low- and middle-permeability layers decreases by 0.3% and 0.1%. In the high-permeability layer, the increase is only 0.1% with an increase of 0.18% in enhancement oil recovery. For Scheme 8, the interfacial tension is 5.0×10^{-4} mN/m, the maximum capillary number of well grid in the oil displacement process is 0.75, and the driving status under the "Type II" driving situation is in a long time and a wide range. From the residual oil distribution data of high-permeability layer grids in the termination time in Table 7.6, the residual oil content of grids drops to zero, and the residual oil content v_{ow} decreases to a certain extent relative to the initial S_{or}^{H} of 13.2%; its lowest value is 8.72%. There are six grids marked in orange on each corner of the two wings of the layer where the number stands for the residual oil content greater than the residual saturation limit S_{or}^{H}. The oil remains in these grids spaces with a maximum value of 1.37%, consisting of three parts: the oil in pure oil space and pure water space, and the flowing oil captured in oil/water space. The residual oil content v_{ow} of the grids is greater than the residual saturation limit S_{or}^{H} (these grids are marked in light green) and the maximum value of v_{ow} is 0.85%. Although the oil in oil/water spaces of some high-permeability layer grids is driven away, enough flowing oil was captured again by a large amount of grid oil/water space, so the driving effect is not ideal. When the displacement process is terminated, the residual oil value in this high-permeability layer decreases by 0.4% compared with Scheme 7, while the lower- and middle-permeability layers decrease by 0.18% and 0.02% respectively. The oil recovery of Scheme 8 increased by 0.23% compared with Scheme 7. The overall evaluation of the four oil displacement schemes shows that when interfacial tension decreases, the capillary number gradually increases in the displacement process as does the recovery percentage. In "Type II" driving conditions, however, there is a special relation between the increased capillary number and the decreased residual oil saturation, and it leads to an asynchronous increasing rate between the recovery percentage and the capillary number. The recovery percentage of Scheme 5 increases by 1.06% compared with Scheme

4 and Scheme 6, 0.75% compared with Scheme 5 and Scheme 7, 0.18% compared with Plan 6 and Scheme 8, and 0.23% compared with Scheme 7. Obviously, increases for Schemes 7 and 8 are very small. The properties of decreasing interfacial tension in underground of surfactant should be related to economic investment. The ultra-low interfacial tension surfactant selected by the two schemes is not desirable because it is necessary to invest quite high costs to produce such high-quality surfactant. The crude oil produced in Schemes 5 and 6 is basically crude oil originally stored in the pure oil space V_o of each layer, which is relatively easy to drive out, and the increasing oil recovery range is more than 30%, and the oil increase for ton equivalent polymer is more than 58 tons. We recommend the surfactant with interfacial tension between 5×10^{-4} and 2.5×10^{-3} mN/m underground in view of the considerable economic benefits.

In Scheme 9, the interfacial tension is 0.00025 mN/m, making it an extra-ultra-low interfacial tension system. As can be seen from Table 7.6, almost all grids of layers are marked in green, indicating that the residual oil on these grid positions was retained in the oil/water space with a minimum v_{ow} of 8.0%. In the edges and corners of the oil layer grids, 12 grids marked in light green number and 4 grids marked in orange have a maximum v_{ow} of 0.99% and 1.38%, respectively, showing a good displacement effect. In Table 7.5, the residual oil content of the three intervals is 28.4%, 16.2%, and 11.8% respectively at the termination moment of Scheme 9. Compared with Scheme 8, the residual oil content in the low- and the high-permeability layers decreased by 0.1% and 1.4% respectively, while it increased by 0.3% in the middle-permeability layer. The enhancement oil recovery percentage reaches 32.29% and increases 0.7% compared with Scheme 8. The parameters of Scheme 10 have been calculated under the condition that the interfacial tension is 2.0×10^{-5} mN/m, meeting the displacement requirement, and the surfactant concentration is 0.3%. As can be seen from Table 7.6, only the residual oil remained in the oil/water space in the high-permeability layer at the termination moment, with a maximum value of 10.96%. In Table 7.5, the residual oil content of the three layers is 28.5%, 13.3%, and 8.47% respectively at the termination moment. Compared with Scheme 8, the residual oil content increases by 0.1% in the low-permeability layer, decreases by 2.9% in the middle-permeability layer, and decreases by 3.33% in the high-permeability layer. The enhancement oil recovery percentage of Scheme 10 increases by 34.30% which increases by 2.01% compared with Scheme 9. Compared with Scheme 6, the residual oil content of Scheme 10 decreases by 0.50%, 2.70%, and 5.03% in the low-, middle-, and high-permeability layers, respectively. The enhancement oil recovery percentage of Scheme 10 increases by 3.12% compared with Scheme 6. In particular, it needs to be pointed out that taking the price of surfactant as 2.5 times that of polymer, the oil increment for per ton equivalent polymer reached 51.42 tons. Furthermore, if the surfactant price dropped to 1.5 times the polymer price, it will be 64.62 tons with the same scheme as the previous, so the use of a high-viscosity ultra-low interfacial tension system is worthy of attention and research.

After deciding the interfacial tension of oil displacement system underground, it is necessary to understand the influence of the changes of surfactant concentration on oil displacement effect when the interfacial tension reaches the standard requirements. In the previous study on Scheme 6 in Table 7.5, we recommended the system with underground interfacial tension of 1.25×10^{-3} mN/m and the lower limit of surfactant concentration of 0.075%, that is to say, if the concentration of surfactant in the system is greater than 0.075%, the underground interfacial tension can reach 1.25×10^{-3} mN/m. The lower limit of surfactant concentration is different, which leads to different oil displacement effect and different oil increment for per ton equivalent polymer for the system meets the standard requirements. The lower limit of concentration of five different surfactant sis calculated, and the scheme results are listed in Table 7.7.

It can be clearly seen from Table 7.7 that the lower limit of surfactant concentration gradually increases, and the remaining oil value S_{or} in the three-layer oil reservoir gradually increases, and the enhancement oil recovery range decreased gradually, with the maximum and minimum difference of 1.2%. Similarly, the oil increment for per ton equivalent polymer gradually decreases, with the maximum and minimum difference of 2.31 tons. The results suggest that the surfactant with the lower concentration limit should be chosen. It also tells us to choose the relatively lowest concentration of surfactant when the lower limit concentration is not very low. Meanwhile, the influence of selection results on oil displacement effect should be recognized. For example, in Table 7.7, we chose the system with the lower limit of the concentration of surfactant of 0.15%, and found that the enhancement oil recovery percentage in Scheme 3 to be 30.38%, which is 0.80% lower than that in Scheme 1. The oil increment for per ton equivalent polymer was 58.28 tons, which is 1.54 tons less than that of Scheme 1. On this basis, we can say that fully understand the performance of the oil displacement system.

7.3.3 Impact of reservoir heterogeneity on oil displacement efficiency

The reservoir heterogeneity of the Placanticline Structure at Daqing Oilfield gradually increased from south to north, which is the basic geologic characteristic of the oil reservoir. Chapter 6 established a digital geological model, she basic data of which given in Table 7.8.

The geological data in Table 7.8 indicate that the reservoir heterogeneity gradually increased from the Xinger block area in the south to Beisanxi block area in the north, and the compound displacement residual oil saturation S_{or}^{H} related to the compound displacement effect also presented regular changes. The residual oil saturation decreased with the increase of reservoir

TABLE 7.7 The effect of five flooding scheme with lower limit concentration and interfacial tension meeting the standard requirements.

Scheme	The lower limit concentration (%)	Implementation time (days)	Reservoir maximum average pressure ratio	Maximum viscosity of displacing phase (mPa s)	Maximum capillary number of the grid in flooding process	Recovery percent (%)	Enhancement of recovery percent (%)	Stratified residual oil S_{or} (%)			Total chemical dosage equivalent polymer (t)	Oil increase for T equivalent polymer (t/t)
								Low permeability layer	*Medium permeability layer*	*High permeability layer*		
1	0.075	1045	1.00	33.03	0.2966	74.88	31.18	29.0	16.0	17.5	665.8	59.82
2	0.10	1034	0.99	32.56	0.2966	74.56	30.86	29.5	16.17	13.67	665.8	59.20
3	0.15	1028	1.01	32.56	0.2965	74.08	30.38	30.25	16.47	13.69	665.8	58.28
4	0.20	1027	1.01	32.56	0.2965	73.83	30.13	30.62	16.67	13.72	665.8	57.80
5	0.25	1028	1.01	32.57	0.2965	73.68	29.98	39.83	16.78	13.75	665.8	**57.51**

TABLE 7.8 Geological model key parameters of industrial combination flooding in the 4th Oil Production Company in Daqing Oilfield.

Scheme	Stratified permeability K ($10^{-3}\mu m^2$)			Bound water saturation S_{wr}^L (%)			Water flood residual oil saturation S_{or} (%)			Saturation of combination flooding residual oil $S_{or}^H(v_{ow})$ (%)		
	Upper layer	*Middle layer*	*Lower layer*	*Upper layer*	*Middle layer*	*Lower layer*	*Upper layer*	*Middle layer*	*Lower layer*	*Upper layer*	*Middle layer*	*Lower layer*
Xinger block	100	215	525	24.0	22.0	21.0	36.5	34.5	32.5	15.50	14.50	13.50
Nanwu block	100	230	600	24.0	22.0	21.0	37.0	36.0	34.5	15.50	14.45	13.35
Beiyi Duandong	100	250	750	24.0	22.0	21.0	38.0	37.0	36.0	15.50	14.40	13.20
Beisanxi block	100	275	825	24.0	22.0	21.0	38.5	38.0	37.0	15.50	14.35	13.05

permeability, which is consistent with the core experiment results in Ref. 4. It should be noted that Table 7.8 does not separately list the initial oil content v_o and v_{ow} in the initial oil amount in oil/water coexistence space. These two important parameters, v_o and v_{ow}, are not difficult to calculate. The initial oil amount v_{ow} is the same as the residual oil saturation value S_{or}^H, and the initial oil content the v_o is equal to $1 - S_{or}^L - S_{or}^H$.

The optimized displacement scheme was calculated based on the geology model of Xinger block in No. 4 Oil Production Plant. The compound system slug was 0.3 PV, the surfactant concentration was 0.3%, the interfacial tension of the system was 1.25×10^{-3} mN/m, subsequent polymer slug was 0.6 PV, the injector-producer spacing was 125 m, and the slug injection velocity was 0.4 PV/y. Based on the pressure limit obtained by fitting the field experiments, the polymer concentration in the slug is adjusted to operate Scheme 1, and the optimized slug polymer concentration obtained is 2030 mg/L. The specific value of the maximum average pressure of the oil reservoir in the scheme and the benchmark pressure is less than 1. The calculation results listed in Table 7.9 show that the enhancement of recovery range of scheme CF1 is 28.34% and the oil increase for ton equivalent polymer is 58.90 tons. The field experiments indicated that the injection capacity increased with the increase of reservoir heterogeneity. For the corresponding blocks of various oil production plants, the scheme in which the slug concentration is 2030 mg/L was calculated, and the calculation results are listed in column "CF1" in Table 7.9. The enhancement of recovery is greater than 28%, and from south to north, the oil increase for ton equivalent polymer is about 60 tons. We take the maximum average reservoir pressure as the common injection pressure limit in the field test scheme of Xingerxi block of the No. 4 Oil Production Plant, and then increase the viscosity of displacement systems to calculate the scheme of all test blocks; the results are listed in column "CF2" in Table 7.9. As shown in the table, from the south to the north, with the increases of polymer concentration of slug and enhancement oil recovery, the enhancement oil recovery can reach 29.49%, 31.18%, and 32.36% respectively in the Nanwu block of the No. 2 Oil Production Plant, the Duangdong of Beiyi block of the No. 1 Oil Production Plant, and the Beisanxi block of the No. 3 Oil Production Plant. The results shows that it is of great significance to study and determine the reservoir pressure limit to improve the recovery efficiency of combination displacement.

Based on the previous schemes, we take the surfactant concentration of 0.3% and interfacial tension of 2×10^{-5} mN/m of the displacement system to calculate the scheme of corresponding high-viscosity ultra-low interfacial tension system; the results are listed in column "CF3" in Table 7.9. As can be seen from Table 7.9, compared with the scheme of "CF2," the relative enhancement oil recovery is 1.08% in the Xingerxi block of the No. 4 Oil Production Plant, 2.98% in the Nanwu block of the No. 2 Oil Production Plant, 3.09% in the Duandong of Beiyi block of the No. 1 Oil Production Plant, and 3.39% in the Beisanxi block of the No. 3 Oil Production Plant. The oil increment for ton equivalent

TABLE 7.9 Optimization scheme results of industrial combination flooding in the 4th Oil Production Company in Daqing Oilfield (well spacing is 125 m).

Experiment	Scheme	Implementation time (days)	Reservoir maximum average pressure ratio	Slug polymer concentration (mg/L)	Maximum viscosity of displacing phase (mPa s)	Recovery percent of reserves (%)	Enhancement of recovery (%)	Stratified residual oil S_{or} (%)			Surfactant consumption (t)	Total chemical dosage equivalent polymer (t)	Oil increase for t equivalent polymer (t/t)
								Upper layer	*Middle layer*	*Lower layer*			
Xinger block	Water flooding				0.6	47.36		45.58	41.24	26.06			
	CF1,2	1055	0.9894	2030	27.26	75.70	28.34	26.13	16.39	15.10	174.1	614.6	58.90
	CF2-1	1573	1.0053	2850	46.45	78.02	30.66	22.44	15.31	13.42	174.1	678.0	57.77
	CF3	1042	0.9939	2050	28.30	76.74	29.42	29.38	15.55	9.28	**174.1**	**792.2**	**47.44**
	CF3-1	1532	0.9967	2850	47.70	80.28	32.92	24.47	13.36	8.15	**174.1**	**852.1**	**49.35**
Nanwu block	Water flooding				0.6	45.56		46.67	42.68	37.50			
	CF1	1054	0.9068	2030	27.24	74.50	28.94	28.78	16.53	14.15	174.1	614.6	60.15
	CF2	1056	0.9570	2104	28.66	75.05	29.49	27.85	16.28	14.09	174.1	627.5	60.03
	CF2-1	1567	0.9970	2950	50.58	77.54	31.98	23.44	15.21	1370	174.1	692.6	58.98
	CF3	1032	0.9930	2160	30.41	78.03	32.47	27.50	14.63	9.07	**174.1**	**811.4**	**51.12**
	CF3-1	1559	1.0055	2975	51.93	80.20	34.64	24.94	13.23	7.99	**174.1**	**870.4**	**50.84**

Beiyi Duandong block	Water flooding				0.6	43.70		48.54	43.95	38.68			
	CF1	1028	0.8079	2030	27.39	73.16	29.46	31.75	16.98	13.81	174.1	614.6	61.23
	CF2	1045	0.9955	2324	33.48	74.88	31.18	29.01	15.98	13.54	174.1	665.8	59.82
	CF2-1	1539	1.0073	3175	58.16	77.09	33.39	25.38	15.09	12.92	174.1	725.5	58.79
	CF3	1073	0.9918	2375	34.46	77.96	34.26	28.47	14.47	8.48	**174.1**	**848.8**	**51.56**
	CF3-1	1552	0.9997	3225	60.87	79.28	35.58	26.91	13.52	7.87	**174.1**	**907.0**	**50.11**
Beisanxi block	Water flooding				0.6	42.42		49.85	44.75	39.56			
	CF1	1020	0.7664	2030	27.28	71.90	29.48	34.88	17.01	13.78	174.1	614.6	61.27
	CF2	1043	0.9727	2395	34.35	74.80	32.38	29.50	15.80	13.58	174.1	678.2	60.99
	CF2-1	1565	0.9990	3280	61.50	76.77	34.35	25.50	14.96	13.35	174.1	758.9	57.82
	CF3	1094	0.9911	2468	36.18	78.19	35.77	28.61	14.03	8.27	**174.1**	**865.0**	**52.82**
	CF3-1	1646	0.9911	3280	61.65	79.50	37.08	26.99	13.05	7.77	**174.1**	**944.5**	**50.15**

CF, combination flooding.

polymer is 47.44 tons in the Xingerxi block, 51.12 tons in the Nanwu block, 51.16 tons in the Duandong of Beiyi block, 52.82 tons in the Beisanxi block, compared with the scheme of "CF2." The oil increment for ton equivalent polymer is deduced significantly, for the price of surfactant was 2.5 times of the price of polymer.

7.3.4 Further optimization of the injection velocity of schemes

In Section 7.3.2, we put forward the measure of "slow down velocity and maintain viscosity" and we recommend injection velocity of 0.40 PV/y when the injection-production interval is 125 m. However, when author came back to Daqing Oilfield to exchange ideas in the Spring of 2018, the injection velocity of slug of the compound system had been reduced to 0.2 PV/y. We immediately calculated and studied to optimize the scheme as follows: the slug volume of the compound system remained unchanged, the polymer slug volume decreased to 0.5 PV, and the injection velocity of the two-stage slug decreased to 0.2 PV/y. The injection velocity is still 0.4 PV/y after the subsequent water injection, and in Table 7.9, the Scheme 2-1 is the calculation result of the re-optimization scheme of each oil production plant. Now let's analyze the re-optimization scheme. The maximum implementation time of the scheme is 1573 days, slightly exceeding the time limit of 1570 days required for the stability of surfactant, which can be considered as meeting the stability requirements. The stability time of surfactant is basically consistent with that of the oil displacement process, which can give full play to the stability effect of the surfactant. Furthermore, the deviation between the maximum pressure in bottom hole and pressure limit during injection is less than 1%, which meets the engineering and system viscosity shear degradation requirements. Then we evaluated the economic and technical effects of each Oil Production Plant. Compared with Scheme 2, the oil recovery percentage of Scheme 2-1 increased by 2.32% in the No. 4 Oil Production Plant, and the residual oil content decreased by 3.69%, 1.08%, and 0.64% respectively in the low-, middle-, and high-permeability layers. In addition, the oil increment for ton equivalent polymer is 57.77 tons for Scheme 2-1 in the No. 4 Oil Production Plant. In the No. 2 Oil Production Plant, the corresponding data are 2.49%, 4.41%, 1.07%, 0.39%, and 58.98 tons respectively. In the No. 1 Oil Production Plant, the corresponding data are 2.21%, 3.63%, 0.89%, 0.62%, and 58.79 tons respectively. In the No. 3 Oil Production Plant, the corresponding data are 1.97%, 4.0%, 0.84%, 0.23%, and 57.82 tons respectively.

The grid residual oil distribution data of each interval at the terminal time of Scheme 2-1 of the four Oil Production Plants are listed in Tables 7.10 and 7.11. According to the data in the two tables, in the low permeability-layer, the percentage of the grids of residual oil content v_o, which is less than 1% in the total grids, is 39.5% in the Xinger block, 38.27% in the Nanwu block, 34.57% in the

TABLE 7.10 The grid residual oil distribution of each intervals at the terminal time of the optimization scheme for some test blocks (%).

Test Block	Intervals	Line/Row	1	2	3	4	5	6	7	8	9
Xinger Block of the 4th Oil Production Company	Low Permeability Layer	1	0.08	0.14	0.23	0.29	0.41	0.76	2.64	22.55	29.18
		2	0.14	0.19	0.24	0.29	0.42	0.76	2.28	20.38	29.48
		3	0.23	0.24	0.25	0.32	0.47	0.83	2.22	19.98	29.59
		4	0.29	0.29	0.32	0.40	0.59	1.00	2.70	20.89	29.75
		5	0.41	0.42	0.47	0.59	0.82	1.32	6.68	23.07	29.75
		6	0.76	0.76	0.83	0.99	1.32	2.55	13.37	24.67	29.52
		7	2.61	2.25	2.23	2.67	6.65	13.41	19.88	25.39	29.67
		8	22.56	20.39	19.91	20.87	23.09	24.72	25.64	27.30	28.91
		9	29.23	29.52	29.62	29.77	29.77	29.52	29.65	28.92	29.63
	Middle Permeability Layer	1	0.05	0.08	0.14	0.19	0.24	0.29	0.43	0.85	6.13
		2	0.08	0.11	0.15	0.19	0.24	0.29	0.42	0.84	3.44
		3	0.14	0.15	0.18	0.20	0.23	0.30	0.43	0.83	3.66
		4	0.19	0.19	0.20	0.21	0.25	0.32	0.46	0.85	3.60
		5	0.24	0.24	0.23	0.25	0.30	0.36	0.53	0.90	3.48
		6	0.29	0.29	0.30	0.32	0.36	0.45	0.61	0.94	3.32
		7	0.43	0.42	0.42	0.46	0.53	0.60	0.69	0.94	3.12
		8	0.85	0.83	0.83	0.85	0.89	0.93	0.93	1.16	2.76
		9	6.04	3.47	3.69	3.64	3.52	3.34	3.16	2.81	20.82
	High Permeability Layer	1	8.24	12.84	0.11	0.13	0.15	0.17	0.19	0.27	0.97
		2	12.84	0.10	0.12	0.13	0.15	0.17	0.18	0.26	0.96
		3	0.11	0.12	0.13	0.15	0.16	0.17	0.18	0.26	0.96
		4	0.13	0.13	0.15	0.15	0.16	0.15	0.18	0.25	0.96
		5	0.15	0.15	0.16	0.16	0.15	0.16	0.19	0.26	0.95
		6	0.17	0.17	0.17	0.15	0.16	0.17	0.19	0.26	0.88
		7	0.19	0.18	0.18	0.18	0.18	0.19	0.21	0.20	0.11
		8	0.27	0.26	0.26	0.25	0.26	0.25	0.19	11.72	9.89
		9	0.95	0.94	0.94	0.93	0.91	0.80	11.98	7.79	8.11
Nanwu Block of the 2th Oil Production Company	Low Permeability Layer	1	0.08	0.13	0.21	0.27	0.41	0.81	4.46	24.65	29.70
		2	0.13	0.17	0.22	0.28	0.42	0.81	3.02	23.11	30.03
		3	0.21	0.22	0.24	0.31	0.48	0.89	3.04	23.16	30.24
		4	0.27	0.28	0.31	0.41	0.61	1.13	7.81	24.21	30.24
		5	0.40	0.42	0.48	0.61	0.88	1.70	11.18	25.36	30.16
		6	0.81	0.81	0.89	1.13	1.69	5.25	18.69	26.11	29.71
		7	4.58	3.00	2.99	7.71	11.18	18.79	23.43	26.63	29.76
		8	24.54	22.96	23.04	24.18	25.38	26.15	26.72	28.20	29.07
		9	29.77	30.06	30.28	30.28	30.20	29.76	29.79	29.09	29.88
	Middle Permeability Layer	1	0.04	0.07	0.12	0.17	0.22	0.27	0.40	0.80	5.82
		2	0.07	0.10	0.13	0.17	0.21	0.27	0.40	0.79	3.29
		3	0.12	0.13	0.15	0.18	0.21	0.28	0.40	0.79	3.50
		4	0.17	0.17	0.18	0.19	0.23	0.30	0.44	0.81	3.44
		5	0.22	0.21	0.21	0.23	0.27	0.34	0.50	0.86	3.31
		6	0.26	0.27	0.28	0.30	0.34	0.43	0.57	0.89	3.14
		7	0.40	0.40	0.40	0.43	0.50	0.57	0.66	0.89	2.96
		8	0.80	0.79	0.79	0.81	0.85	0.89	0.89	1.13	2.66
		9	5.69	3.28	3.50	3.44	3.30	3.12	2.96	2.68	22.00
	High Permeability Layer	1	7.84	12.47	0.09	0.11	0.13	0.15	0.16	0.24	0.83
		2	12.47	0.09	0.10	0.11	0.13	0.15	0.16	0.23	0.83
		3	0.09	0.10	0.11	0.12	0.13	0.14	0.16	0.22	0.82
		4	0.11	0.11	0.12	0.13	0.14	0.13	0.16	0.22	0.82
		5	0.13	0.13	0.13	0.14	0.13	0.13	0.16	0.23	0.81
		6	0.15	0.15	0.14	0.13	0.13	0.15	0.16	0.23	0.74
		7	0.16	0.15	0.16	0.16	0.16	0.16	0.18	0.20	0.02
		8	0.23	0.22	0.22	0.22	0.22	0.22	0.18	11.69	9.61
		9	0.82	0.81	0.80	0.80	0.77	0.74	11.79	7.43	7.81

Duandong of the Beiyi block, and 32.10% in the Beisanxi block. For the grids in which v_o is between 1% and 5%, the corresponding percentages are 14.81%, 12.35%, 8.64%, and 8.64% respectively. For the grids in which v_o is between 5% and 10%, the corresponding percentages are 2.47%, 3.70%, 7.41%, and 4.94% respectively. For the grids that have v_o between 10% and 20%, the corresponding percentages are 6.17%, 4.94%, 6.17%, and 8.64% respectively. For the grids that have v_o greater than 20%, the corresponding percentages are 37.04%, 40.74%, 43.21%, and 45.68% respectively. It can be seen that from south to north, the range of low residual oil saturation gradually decreased, while the range of high value gradually increased, and the displacement effect gradually became worse. In middle-permeability layer, the percentage of the grids of residual oil content v_o which is less than 1% in the

TABLE 7.11 The grid residual oil distribution of each intervals at the terminal time of the optimization scheme for some test blocks (%).

Test Blocks	Intervals	Line/Row	1	2	3	4	5	6	7	8	9
Duandong of the Beiyi block in the 1th Oil Production Company	Low Permeability Layer	1	0.07	0.11	0.18	0.25	0.40	0.90	8.17	25.46	30.53
		2	0.11	0.15	0.19	0.25	0.41	0.94	7.52	26.62	30.89
		3	0.18	0.19	0.22	0.29	0.47	1.08	9.22	26.30	30.84
		4	0.25	0.25	0.29	0.40	0.63	1.41	12.77	26.29	30.56
		5	0.40	0.41	0.47	0.63	1.02	3.66	19.35	27.01	30.30
		6	0.90	0.94	1.09	1.41	3.51	14.54	23.03	27.36	29.85
		7	8.25	7.62	9.25	12.84	19.34	23.17	25.91	27.81	29.52
		8	25.45	26.52	26.23	26.24	26.98	27.37	27.82	28.17	31.02
		9	30.73	31.01	30.97	30.74	30.48	29.95	29.92	29.30	29.87
	Middle Permeability Layer	1	0.03	0.06	0.10	0.14	0.18	0.23	0.34	0.70	4.72
		2	0.06	0.08	0.11	0.15	0.17	0.23	0.34	0.70	2.96
		3	0.10	0.11	0.13	0.15	0.18	0.24	0.35	0.70	3.20
		4	0.14	0.15	0.15	0.16	0.20	0.26	0.38	0.72	3.13
		5	0.18	0.17	0.18	0.20	0.24	0.29	0.43	0.75	2.90
		6	0.23	0.23	0.24	0.26	0.29	0.37	0.50	0.78	2.67
		7	0.34	0.34	0.35	0.38	0.43	0.50	0.57	0.78	2.49
		8	0.70	0.69	0.69	0.71	0.75	0.77	0.78	0.88	2.05
		9	4.24	2.82	3.10	3.08	2.95	2.75	2.54	2.28	22.60
	High Permeability Layer	1	6.932	11.31	0.08	0.08	0.09	0.11	0.12	0.17	0.62
		2	11.31	0.10	0.10	0.10	0.10	0.11	0.11	0.16	0.61
		3	0.08	0.10	0.10	0.11	0.11	0.11	0.12	0.16	0.61
		4	0.08	0.10	0.11	0.11	0.11	0.10	0.12	0.16	0.61
		5	0.09	0.10	0.11	0.11	0.10	0.11	0.12	0.17	0.59
		6	0.11	0.11	0.11	0.10	0.11	0.12	0.14	0.22	0.61
		7	0.12	0.11	0.12	0.12	0.13	0.13	0.17	12.50	11.57
		8	0.17	0.16	0.16	0.16	0.17	0.19	12.45	10.23	8.20
		9	0.61	0.60	0.60	0.59	0.63	0.59	9.96	6.29	6.58
Beisanxi Block in the 3th Oil Production Company	Low Permeability Layer	1	0.07	0.11	0.18	0.25	0.41	0.98	12.53	25.65	31.09
		2	0.11	0.15	0.19	0.26	0.42	1.01	9.27	26.00	31.36
		3	0.18	0.19	0.22	0.30	0.49	1.16	12.12	27.88	31.35
		4	0.25	0.26	0.30	0.41	0.65	1.71	16.93	27.56	31.09
		5	0.41	0.42	0.49	0.65	1.14	7.44	21.96	27.78	30.69
		6	0.97	1.00	1.16	1.71	7.46	17.74	24.14	28.01	30.14
		7	12.41	9.09	12.14	16.98	21.95	24.14	27.04	28.55	30.07
		8	25.65	26.00	27.85	27.56	27.80	28.02	28.58	29.49	29.60
		9	31.13	31.37	31.35	31.08	30.67	30.14	30.07	29.60	30.42
	Middle Permeability Layer	1	0.15	0.11	0.12	0.14	0.17	0.21	0.31	0.62	3.64
		2	0.11	0.12	0.12	0.14	0.17	0.22	0.32	0.62	2.61
		3	0.12	0.12	0.13	0.14	0.17	0.22	0.32	0.62	2.84
		4	0.14	0.14	0.14	0.16	0.19	0.24	0.35	0.64	2.83
		5	0.17	0.17	0.17	0.19	0.22	0.27	0.39	0.67	2.74
		6	0.21	0.22	0.22	0.24	0.27	0.33	0.44	0.69	2.57
		7	0.31	0.31	0.32	0.35	0.39	0.44	0.50	0.69	2.37
		8	0.61	0.61	0.62	0.63	0.66	0.68	0.69	0.94	2.15
		9	3.59	2.62	2.84	2.83	2.73	2.56	2.38	2.17	22.98
	High Permeability Layer	1	6.77	11.14	0.26	0.16	0.12	0.11	0.11	0.16	0.56
		2	11.14	0.10	0.18	0.13	0.11	0.11	0.11	0.15	0.55
		3	0.27	0.18	0.12	0.11	0.11	0.11	0.11	0.15	0.55
		4	0.16	0.13	0.11	0.10	0.11	0.09	0.11	0.15	0.55
		5	0.12	0.11	0.11	0.11	0.09	0.10	0.12	0.19	0.59
		6	0.11	0.11	0.11	0.09	0.10	0.11	0.14	0.22	0.80
		7	0.11	0.11	0.11	0.11	0.12	0.13	0.16	12.04	11.76
		8	0.16	0.15	0.15	0.15	0.20	0.19	11.97	9.77	7.69
		9	0.55	0.54	0.54	0.53	0.64	0.60	9.35	5.86	6.11

total grids, is 77.78% in the Xinger block and the Nanwu block, and 79.01% in the Duandong of the Beiyi block and the Beisanxi block. For the grids that have v_o between 1% and 5%, the corresponding percentages are 18.52%, 18.52%, 19.75%, and 19.75% respectively. For the grids that have v_o between 5% and 10%, the corresponding percentages are 2.47%, 2.47%, 7.41%, and 4.94% respectively. For the grids that have v_o between 10% and 20%, the corresponding percentages are all zero. For the grids that have v_o greater than 20%, there are only oil wells grids in the four blocks, the residual oil content v_o is 20.82%, 22.00%, 22.60%, and 22.98% respectively. The displacement effect is relatively improved from south to north. In the high-permeability layer, it should be noted that the remaining oil saturation of the grids near water wells and oil wells in the two tables falls below the limit of residual oil saturation S_{or}^{H}, so the grid value of v_{ow} is marked in green, and the other grid values of v_{ow} are equal to $S_{or} - S_{or}^{H}$. Through numerical analysis, we find that these values are relatively small, and it is difficult to identify whether there is residual oil in the pure oil space V_o and pure water space V_w near the grids marked in green. In view of this situation, the grid values are marked in red. We analyzed the data and found that the proportion of the grids of residual oil content v_o less than 1% in the total grids is 90.12% in the Xinger block and the Nanwu block, and 79.01% in the Duandong of the Beiyi block and the Beisanxi block. For the others grids, the residual oil content v_{ow} higher than 1%, which is marked in green, the minimum value of v_{ow} is 7.79% in the Xinger block, 7.43% in the Nanwu block, 6.29% in the Duandong of the Beiyi block, and 5.86% in the Beisanxi block. It is clear that the displacement effect is further improved from south to north.

It can be seen from the research that the main target of oil recovery via adoption of a high-viscosity and ultra-low interfacial tension system is the crude oil retained in the microscopic pure oil space V_o and pure water space V_w of reservoirs. This part of crude oil is relatively easy to produce, and good technical and economic effects can be obtained by adopting the optimized oil displacement scheme proposed.

The same optimization measures are taken for Scheme 3 of the compound displacement system with high-viscosity extra-ultra-low interfacial tension, so the slug volume is 0.3 PV, the polymer slug volume is 0.5 PV, the injection velocity of the two-stage slug is reduced to 0.2 PV/y, and the injection velocity of the subsequent transfer to clear water is 0.4 PV/y. In Table 7.9, Scheme 3-1 is the calculation result of the re-optimization scheme of compound displacement. The maximum implementation time of the scheme is 1646 days, exceeding the time limit by less than 5%, and meeting the stability requirements of surfactant. Furthermore, the deviation between the maximum pressure in bottom hole and pressure limit during injection is less than 1%, which meets the engineering and system viscosity shear degradation requirements. For the oil layers of the Xingerxi block, Scheme 3-1 is compared with Scheme 3. The enhancement oil recovery will increase by 4.91% in the low-permeability layer, 2.19% in the middle-permeability layer, and 1.13% in the high-permeability layer. The oil recovery percentage will increase by 3.50%, and the oil increment for ton equivalent polymer will decrease by 1.91 tons. In Scheme 3-1, compared with Scheme 2-1, the enhancement oil recovery decreases by 2.03% in the low-permeability layer, increases by 1.95% in the middle-permeability layer, and increases by 5.27% in the high-permeability layer. The oil recovery percentage increases by 2.26% and the oil increment for ton equivalent polymer decreases by 8.42 tons. It should be pointed out that the surfactant with high price is used in Scheme 3-1. For the oil layers of the Nanwu block, Scheme 3-1 is compared with Scheme 3. The enhancement oil recovery increases by 2.56% in the low-permeability layer, 1.40% in the middle-permeability layer, and 1.80% in the high-permeability layer. The oil recovery percentage increases by 2.17% and the oil increment for ton equivalent polymer decreases by 0.28 tons. In Scheme 3-1, as compared with Scheme 2-1, the enhancement oil recovery decreases by 1.50% in the low-permeability layer, increases by 1.98% in the middle-permeability layer, and increases by 5.71% in the high-permeability layer. The oil recovery percentage increases by 2.66% and the oil increment for ton equivalent polymer decreases by 8.14 tons. For the oil layers of the Duandong of Beiyi block, Scheme 3-1 is compared with the Scheme 3. The enhancement oil recovery increases by 1.56% in the low-permeability layer, 0.95% in the middle-permeability layer, and 0.61% in the high-permeability layer. The oil recovery percentage increases by 1.32% and the oil increment for ton equivalent polymer decreases by 1.45 tons. In Scheme 3-1, as compared with Scheme 2-1, the enhancement oil recovery decreases by 1.53% in the low-permeability layer, increases by 1.57% in the middle-permeability layer, and increases by 5.05% in the high-permeability layer. The oil recovery percentage increases by 2.19% and the oil increment for ton equivalent polymer decreases by 8.68 tons. For the oil layers of the Beisanxi block, Scheme 3-1 is compared with Scheme 3. The enhancement oil recovery increases by 1.62% in the low-permeability layer, 0.98% in the middle-permeability layer, and 0.50% in the high-permeability layer. The oil recovery percentage increases by 1.31% and the oil increment for ton equivalent polymer decreases by 2.67 tons. In Scheme 3-1, as compared with Scheme 2-1, the enhancement oil recovery decreases by 1.49% in the low-permeability layer, increases by 1.91% in the middle-permeability layer, and increases by 5.58% in the high-permeability layer. The oil recovery percentage increases by 2.73% and the oil increment for ton equivalent polymer decreases by 7.67 tons. It can be clearly seen from this data that in Scheme 3-1, compared with Scheme 3, the oil recovery percentage of each interval increases relatively, and the oil recovery increment amplitude of the low-permeability interval is the largest, while the high-permeability interval is the smallest. Among the blocks, the oil recovery increment amplitude of the southern oil layer is relatively high compared with homogeneous oil layers. In Scheme 3-1, as compared with Scheme 2-1, the oil recovery percentage of the low-permeability layer is relatively reduced, which is obvious in the test of the Xingerxi of block of the No. 4 Oil Production Plant. However, the oil recovery percentage increases relatively in the middle- and high-permeability layers, the oil recovery increment amplitude is prominent with the high-permeability layer, and the enhancement oil recovery of each scheme of these four blocks is greater than 2% compared with Scheme 3.

In order to have a deeper and more accurate understanding of the oil displacement effect of the high-viscosity extra-ultra-low interfacial tension system, the grid residual oil distribution data of each interval at the terminal time of test schemes in Duangdong of the Beiyi block and the Beisanxi block are listed in Table 7.12.

TABLE 7.12 The grid residual oil distribution of each intervals at the terminal time of the scheme with high viscosity extra-ultra-low interfacial tension system for some test blocks (%).

Test Block	Intervals	Line/Row	1	2	3	4	5	6	7	8	9
Duandong of the Beiyi block in the 1th Oil Production Company	Low permeability layer	1	13.89	14.77	15.11	1.27	4.66	11.09	7.36	26.10	30.80
		2	14.77	15.18	15.38	1.58	6.20	10.54	6.77	25.02	30.62
		3	15.11	15.39	1.16	3.64	11.50	7.75	5.39	24.15	30.39
		4	1.28	1.58	3.64	12.24	8.56	4.833	7.93	24.21	30.06
		5	4.74	6.26	11.54	8.54	5.28	2.56	14.03	24.52	29.39
		6	10.14	10.50	7.72	4.72	2.48	8.13	19.92	24.71	30.04
		7	7.22	6.66	5.37	8.17	14.39	20.46	23.36	25.06	29.14
		8	26.07	25.09	24.33	24.47	24.86	25.17	25.54	25.84	28.31
		9	30.85	30.69	30.45	30.07	29.40	28.93	28.57	29.03	29.04
	Middle permeability layer	1	8.22	9.86	9.46	10.37	11.25	12.17	13.18	2.333	4.51
		2	9.86	10.36	9.67	10.46	11.18	11.97	12.75	0.87	4.48
		3	9.46	9.67	10.22	10.76	11.19	11.39	12.27	14.27	4.59
		4	10.36	10.46	10.76	10.96	11.22	11.04	11.78	13.93	6.13
		5	11.24	11.18	11.19	11.21	10.76	10.92	11.49	14.10	9.11
		6	12.15	11.96	11.39	11.03	10.92	10.93	11.43	14.31	8.30
		7	13.15	12.72	12.26	11.76	11.48	11.44	12.35	0.68	9.14
		8	2.33	0.92	14.32	13.72	13.78	14.20	0.96	3.15	6.81
		9	4.52	4.49	4.59	6.96	8.96	8.34	7.55	7.10	19.27
	High permeability layer	1	5.76	8.18	7.45	7.36	7.41	7.50	7.60	8.10	9.78
		2	8.18	8.54	7.65	7.24	7.35	7.44	7.53	7.84	9.18
		3	7.46	7.65	7.24	7.28	7.29	7.37	7.45	7.59	9.01
		4	7.34	7.24	7.28	7.26	7.23	7.28	7.38	7.41	8.89
		5	7.41	7.35	7.30	7.21	7.25	7.27	7.26	7.30	8.88
		6	7.50	7.43	7.37	7.28	7.26	7.25	7.16	7.40	8.85
		7	7.60	7.52	7.44	7.38	7.26	7.16	7.09	7.16	9.50
		8	8.08	7.83	7.57	7.40	7.32	7.27	7.29	7.17	10.53
		9	9.70	9.11	8.91	8.79	8.80	9.71	10.62	0.65	1.42
The Beisanxi block in the 3th Oil Production Company	Low permeability layer	1	14.01	15.06	15.26	1.38	5.79	11.48	8.37	26.98	31.39
		2	15.06	15.20	0.27	1.83	8.25	9.69	6.66	26.24	31.31
		3	15.26	0.08	1.26	4.57	15.07	5.71	7.33	25.92	30.92
		4	1.37	1.82	4.57	15.41	6.54	1.71	10.96	26.13	29.95
		5	5.75	8.29	15.05	6.54	2.06	2.87	18.16	26.42	29.46
		6	11.37	9.67	5.69	1.71	2.61	12.61	22.97	27.03	30.20
		7	8.46	6.72	7.36	11.62	18.40	22.98	25.65	27.38	29.18
		8	27.02	26.33	26.03	26.35	26.66	27.16	27.42	27.84	28.90
		9	31.39	31.30	30.81	29.86	29.40	29.30	29.23	30.36	29.95
	Middle permeability layer	1	8.17	9.77	9.39	10.21	11.14	12.07	13.13	2.33	5.01
		2	9.77	10.36	9.55	10.29	11.08	11.91	12.63	14.18	5.34
		3	9.39	9.55	10.14	10.65	11.09	11.33	12.12	14.31	10.53
		4	10.21	10.30	10.66	10.86	11.23	10.96	11.68	13.92	2.69
		5	11.11	11.08	11.09	11.23	10.70	10.85	11.32	14.18	7.97
		6	12.04	11.91	11.33	10.96	10.84	10.82	11.21	13.77	11.85
		7	13.08	12.60	12.13	11.68	11.32	11.22	12.05	0.28	10.63
		8	2.31	2.80	14.40	13.72	14.07	13.85	0.43	3.26	6.75
		9	4.97	5.28	10.78	2.03	9.13	9.40	9.02	6.09	20.15
	High permeability layer	1	6.77	8.12	7.23	7.27	7.30	7.40	7.51	7.99	10.54
		2	8.12	8.51	7.66	7.17	7.25	7.33	7.43	7.73	9.99
		3	7.24	7.67	7.15	7.17	7.20	7.27	7.35	7.49	9.87
		4	7.24	7.17	7.17	7.16	7.14	7.19	7.29	7.34	9.90
		5	7.30	7.24	7.19	7.14	7.15	7.17	7.18	7.22	9.88
		6	7.39	7.33	7.26	7.18	7.17	7.16	7.08	6.78	10.00
		7	7.50	7.42	7.34	7.28	7.18	7.07	7.01	7.04	9.85
		8	7.98	7.72	7.49	7.35	7.32	7.67	7.10	6.94	9.18
		9	10.29	9.85	9.66	9.63	9.80	9.68	9.40	8.52	9.68

For Scheme 3-1 of the Duandong of the Beiyi, we first analyze the data of the low-permeability layer. Eight grids near the water well are marked in green. There the residual oil content v_{ow} of the grid is less than the limit residual oil saturation S_{or}^H, which is 15.5%, and the maximum v_{ow} is 15.39%. Along the main stream from water well to oil well, the residual oil content v_{ow} of the front 11 grids is less than the limit residual oil saturation S_{or}^H, marked in light green, and grid value is equal to $v_{ow} - S_{or}^H$, with the maximum value of 6.26%. Toward the oil well, there is a "fuzzy" region with 14 grids. In this region, the residual oil saturation v_{ow} includes not only the oil in pure space V_w and pure space V_o, but also the unactivated oil in the oil/water co-existence space, and the maximum grid value is 12.24%. Further on the near the side of the oil well, there is a region marked in red. In this region, the oil in the oil/water space has not been activated, the residual oil content v_o with having a minimum value of 2.48%. Nevertheless, the maximum value of the edge and corner of the layer is 30.85%, and the grid value of the well is 29.04. Then, we analyze the data of the middle-permeability layer. There are 57 grids marked in green with a minimum grid value of 8.22%, and another 15 grids marked in light green, 8 grids marked in orange with maximum value of 9.11%, and 1 grid marked in red with grid value of 19.27%, which represent the residual oil content v_{ow} of the oil well. However, there are just a region marked with green number with a minimum grid value of 5.76%. Compared with the data in Table 7.11, there is a region marked in green on the side of oil well where the residual oil saturation is relatively low in the low-permeability layer. Further on toward the oil well, there are regions marked in light green and orange where the residual oil saturation is slightly greater, which is due to the capture of flowing oil in the oil/water space, and the side of the oil well has relatively low residual oil saturation. The middle-permeability layer is similar to the low-permeability layer, except there is a big region marked in green. In the vicinity of the oil well, there are three grids marked in red, and two grids on the side of the oil well grid have relatively high residual oil saturation, which means the residue of the "oil wall" accumulated during the displacement process. In Scheme 3-1, only a few grids near water wells and oil wells are marked in green; all of them have relatively higher residual oil v_{ow}. Furthermore, it can be seen from Scheme 3-1 of the Beisanxi block in Table 7.12 that the residual oil distribution in the three layers is similar to that in the Duandong of the Beiyi block.

Through analysis and research, we can see the distribution of residual oil in each layer of the high-viscosity ultra-low interfacial tension system displacement. We also have a deeper understanding of the reasons why the oil displacement effect of the high-viscosity ultra-low interfacial tension system is relative better. The compound displacement target of high-viscosity ultra-low interfacial tension system has deepened into the oil/water space with smaller pore radius, and this part of the oil is harder to drive out. By adopting the optimized oil displacement scheme, oil recovery should be increased by another 2%, which is worthy of attention.

7.4 Our hopes for Daqing Oilfield

The classic capillary number experimental curve is the theoretical basis of compound displacement technology. The capillary number experimental curve *QL* was made to perfect the capillary number theory. The establishment of oil/water distribution model in micro space not only deepened the understanding of capillary number curve but also to constitute of microcosmic space and the situations of oil/water distribution in the reservoir. On the basis of these two theoretical innovations, a new permeability curve of compound displacement was drawn from overseas research results. The key parameters of the permeability curve of the compound displacement phase were obtained by experiments, and the compound displacement software IMCFS was developed. The "digital displacement test" research method was created via studies of the field tests at Daqing Oilfield. Through thousands of digital displacement tests, the "sensitivity" of the factors in the displacement scheme were studied, and the optimal slug volume and surfactant concentration were determined to be 0.3 PV and 0.3%, respectively. In order to ensure that the compound slug will play a full role underground, the subsequent polymer slug is integral. The compound slug and polymer slug should have the same polymer concentration value as determined by safe injection requirements; the optimal polymer slug volume is 0.5 PV.

One of our hopes for the Daqing Oilfield is that its enhancement oil recovery increases by more than 30% and that its oil increment for ton equivalent polymer reaches about 60 tons. The scheme 2-1 of every test block in Table 7.9 is the optimal design basis of the compound flooding scheme. This optimal Scheme 2-1 has a clear goal of oil recovery, and it is hardly difficult to produce as much crude oil as possible from the micro pure oil space and pure water space in the oil reservoir. We just need to further optimize the current application scheme, adjust the slug structure, surfactant, and polymer concentration in the slug, and select the right interfacial tension. It should also be noted that compound displacement technology is a "high-quality" displacement technology, and Daqing Oilfield is a "precious place" to implement it. Those blocks that are currently implementing and preparing to implement polymer displacement should be converted to compound displacement as soon as possible. On January 19, 2017, China petroleum news center reported that tertiary oil recovery has been widely used in the first- and second-class oil reservoirs of Daqing Oilfield, with an enhanced recovery percent of 10%. In 2016, polymer displacement produced more than 8.5 million tons of oil, the oil increment for per ton polymer reached 47.3 tons, the compound displacement oil production exceeded 4 million tons, and oil recovery increased by 20%. The produced crude oil is an achievement, and the costs paid include not only the input in the development process, but also the value of the remaining crude oil in the ground, which may be "abandoned" there forever. Spending the least amount to obtain the greatest benefit is the crucial problem that scheme studies make.

We also hope the compound displacement target of a high-viscosity ultra-low interfacial tension system to deepen into the oil/water space with smaller pore radius, and this part of the oil is harder to drive out. By adopting the optimized oil displacement scheme, the oil recovery will increase by 2% again. We hope that a small test can be carried out at both the No. 3 Oil Production Plant and No. 1 Oil Production Plant so as to provide an experimental basis for digital research. It is not too difficult to carry out such an experiment. You need only to select a surfactant with interfacial tension of 2×10^{-5} mN/m under the underground condition of oil and water and a surfactant concentration of 0.3%. If the experiments are successful, the technology can be affirmed. Even if the experiments do not achieve the expected effect of Scheme 3-1, it would still be better than the expected effect of Scheme 2-1. With the experimental results, in-depth digital research can be carried out.

7.5 Conclusions

(1) The two important theoretical research results of the capillary number experiment curve *QL* and micro oil/water distribution platform have laid the foundation for digital displacement test research, and the development of high-level software has created an effective research tool for digital oil displacement test research.
(2) It is a necessary condition for digital displacement test research to carry out a field test. This includes using the software describing the compound displacement mechanism more accurately, taking the geological structure model matching with the software, fitting the field experiment and digitizing the field test, and establishing the digital geological model of compound displacement to describe the main geological characteristics of the reservoir and the related information of compound displacement. On the digital geological model, we can calculate the displacement scheme: digital displacement test. "Digital oil displacement test" is an effective research method for in-depth study of field tests and optimization of oil displacement schemes.
(3) Great achievements have been made in the field application of compound displacement technology at Daqing Oilfield. Using the research methods of digitalized displacement experiments and optimizing and researching the industrial application schemes at Daqing Oilfield, we obtained a number of optimization factors of the oil displacement scheme and further improved the technical and economic indexes of compound displacement.
(4) Under the condition of the surfactant applied at present, we hope that Daqing Oilfield will optimize its oil displacement scheme and ensure that the underground interfacial tension during displacement process is less than 1.25×10^{-3} mN/m. If this is done, the recovery percentage would be increased by more than 30% and the oil increment for ton equivalent polymer would be more than 50 tons. We also hope that Daqing Oilfield will choose high-quality surfactant from home rather than abroad, with interfacial tension of 2×10^{-5} mN/m and high viscosity, then carry out small-scale field tests to understand the oil displacement effect of a high-viscosity, ultra-low interfacial tension system.

Symbol descriptions

μ_w	displacing phase viscosity, mPa s
σ_{ow}	interfacial tension between displaced phase and displaced phase, mN/m
N_c	capillary number

N_{cc}	limit capillary number when residual oil begins to flow after water displacement and polymer displacement, that is, the limit capillary number of oil flow in pure oil subspace V_{o2} initiated by compound displacement
N_{ct1}	limit capillary number when the driving condition changes in the process of compound oil driving, that is, the limit capillary number of oil/water flow in oil/water coexistence space initiated by compound displacement
N_{ct2}	the capillary number of the compound displacement out crude oil in "pure oil" pore space V_o under the condition of "type I" driving status, that, is the limit capillary number when the residual oil saturation no longer decreases under the condition of "type I" driving status
r_o	the pore radius of the interface between the microscopic "pure oil" pore space V_o and the microscopic "pure water" pore space V_w
r_{oc}	pore radius of interface between subspaces V_{o1} and V_{o2} of "pure oil" pore space V_o
r_w	pore radius of the interface between "pure water" pore space V_w and "oil/water coexistence" space
S_o	oil phase saturation
S_{or}	residual oil saturation corresponding to capillary number N_c
S_{or}^H	in the "Type I" driving conditions, minimum residual oil saturation in the displacement process, that is, the residual oil saturation corresponding to the capillary number between the limit capillary number N_{ct2} and N_{ct1}
S_{or}^L	under the condition of low capillary number, when the capillary number N_c is less than or equal to N_{cc}, the limit value of residual oil saturation is equal to the oil content v_{o1} in the subspace V_{o1} of "pure oil" pore space V_o
S_w	water phase saturation
S_{wr}^L	irreducible water saturation
T_1, T_2	displacing condition conversion parameter, which can be obtained by experiment or experimental fit
"Type I" driving status	displacement under the condition that the capillary number is less than or equal to the limit capillary number N_{ct1}
"Type II" driving status	displacement under the condition that the capillary number is higher than the limit capillary number N_{ct1}
v	displacement speed (m/s)
v_o	pore volume percentage of initial oil content in microscopic pore space V_o (Note: abbreviated "pore volume percentage of oil content" is "oil content"), after reservoir

	development, v_o volume includes oil volume retained in space V_o and space V_w
V_o	the pore space with a micro pore radius greater than r_o is initially "pure oil" pore space
v_{o1}	pore volume percentage of initial oil content in microscopic pore space V_{o1}
v_{o2}	pore volume percentage of initial oil content in microscopic pore space V_{o2}
V_{o1}	the "pure oil" pore space V_o with pore radius greater than the r_{oc}
V_{o2}	the "pure oil" pore space V_o with pore radius less than the r_{oc}
v_{ow}	pore volume percentage of initial oil content in microscopic pore space V_{ow}, after reservoir development, v_{ow} volume includes oil volume retained in space V_{ow} and space V_{wo}
V_{ow}	the "pure oil" pore space with pore radius r is smaller than r_w
V_w	"pure water" pore space with the pore radius r from r_o to r_w
V_{wo}	the "pure water" pore space with pore radius r is smaller than r_w
V_k	reservoir heterogeneity coefficient of variation

References

1. Moore TF, Slobod RC. The effect of viscosity and capillarity on the displacement of oil by water. *Producers Monthly*. 1956;8.
2. Taber JJ. Dynamic and static forces required to remove a discontinuous oil phase from porous media containing both oil and water. *Soc Pet Eng J*. 1969;9:3.
3. Foster WR. *A Low Tension Water Displacement Process Employing a Petroleum Sulfonate, Inorganic Salts, and a Biopolymer. SPE 3803;* 1972.
4. Qi LQ, Liu ZZ, Yang CZ, et al. Supplement and optimization of classical number experimental curve for enhanced oil recovery by combination displacement. *Sci China Technol Sci*. 2014;57:2190–2203.
5. Fenglan W, Xiaolin W, Guangyu C, et al. Technical progress of alkaline-surfactant-polymer displacement (ASP) in daqing oilfield. *Pet Geol Oilfield Dev Daqing Oilfield*. 2009;28(5):154–162.
6. Demin W, Jiecheng C, Junzheng W, et al. *Summary of ASP Pilots in Daqing Oilfield. SPE 57288;* 1999.
7. Qi L. *Numerical Simulation Research of Polymer Displacement Engineering*. Beijing: Petroleum Industry Press; 1998.

Index

Note: Page numbers followed by *f* indicate figures and *t* indicate tables.

A

B

C

D

E

F

G

H

I

L

M

X